Thermodynamic Modeling of Solid Phases

Chemical Thermodynamics Set

coordinated by
Michel Soustelle

Volume 3

Thermodynamic Modeling of Solid Phases

Michel Soustelle

iSTE

WILEY

First published 2015 in Great Britain and the United States by ISTE Ltd and John Wiley & Sons, Inc.

ISTE Ltd
27-37 St George's Road
London SW19 4EU
UK

www.iste.co.uk

John Wiley & Sons, Inc.
111 River Street
Hoboken, NJ 07030
USA

www.wiley.com

Library of Congress Control Number: 2015944961

British Library Cataloguing-in-Publication Data
A CIP record for this book is available from the British Library
ISBN 978-1-84821-866-6

Contents

Preface

This book – an in-depth examination of chemical thermodynamics – is written for an audience of engineering undergraduates and Masters students in the disciplines of chemistry, physical chemistry, process engineering, materials, etc., and doctoral candidates in those disciplines. It will also be useful for researchers at fundamental- or applied-research labs dealing with issues in thermodynamics during the course of their work.

These audiences will, during their undergraduate degree, have received a grounding in general thermodynamics and chemical thermodynamics, which all science students are normally taught, and will therefore be familiar with the fundamentals, such as the principles and the basic functions of thermodynamics, and the handling of phase and chemical equilibrium states, essentially in an ideal medium, usually for fluid phases, in the absence of electrical fields and independently of any surface effects.

This set of books, which is positioned somewhere between an introduction to the subject and a research paper, offers a detailed examination of chemical thermodynamics that is necessary in the various disciplines relating to chemical or material sciences. It lays the groundwork necessary for students to go and read specialized publications in their different areas. It constitutes a series of reference books that touch on all of the concepts and methods. It discusses both scales of modeling: microscopic (by statistical thermodynamics) and macroscopic, and illustrates the link between them at every step. These models are then used in the study of solid, liquid and gaseous phases, either of pure substances or comprising several components.

The various volumes of the set will deal with the following topics:

– phase modeling tools: application to gases;

– modeling of liquid phases;

– modeling of solid phases;

– chemical equilibrium states;

– phase transformations;

– electrolytes and electrochemical thermodynamics;

– thermodynamics of surfaces, capillary systems and phases of small dimensions.

Appendices in each volume give an introduction to the general methods used in the text, and offer additional mathematical tools and some data.

This series owes a great deal to the feedback, comments and questions from all my students are the *Ecole nationale supérieure des mines* (engineering school) in Saint Etienne who have "endured" my lecturing in thermodynamics for many years. I am very grateful to them, and also thank them for their stimulating attitude. This work is also the fruit of numerous discussions with colleagues who teach thermodynamics in the largest establishments – particularly in the context of the group "Thermodic", founded by Marc Onillion. My thanks go to all of them for their contributions and conviviality.

This volume of the set is devoted to modeling of solid phases.

Chapter 1 discusses the modeling of pure solids. Oscillator models (Einstein's and Debye's) are used to calculate canonical partition functions for four types of solid: atomic, ionic, molecular and metallic. These canonical partition functions can be employed, first to calculate the specific heat capacities at constant volume, and second to determine the expansion coefficients with the Grüneisen parameters.

Chapter 2 looks at the modeling and characterization of solid solutions. Following a qualitative description of the different types of solid solution of substitution and insertion, the short-distance and long-distance order coefficients are introduced. Simple solution models are briefly described and the thermodynamics of the order/disorder transformations in alloys is presented. The chapter ends with the experimental determination of the activity coefficients of the components of a solid solution.

The third chapter deals with non-stoichiometry of solids, and therefore point defects in pure solids. Equilibria between defects are discussed in the context of quasi-chemical phenomena.

The fourth and final chapter looks at the question of point defects in solid solutions that are slightly or highly concentrated. The role of doping of insulating and semiconductive ionic materials is discussed, as is the description of models of gas dissolution in solids. The chapter finishes with an examination of the methods used to calculate the equilibrium constants of point-defect creation.

Two appendices are given at the end of the book, discussing the Lagrange multiplier method and a method for solving Schrödinger's equation.

Michel SOUSTELLE
Saint-Vallier,
July 2015

Notations and Symbols

$\{gas\}$ pure, $\{\{gas\}\}$ in a mixture, $(liquid)$ pure, $((liquid))$ in solution, $\langle solid\rangle$ pure, $\langle\langle solid\rangle\rangle$ in solution

A:	area of a surface or an interface.
$A_H^{(12)}$:	Hamaker constant between two media 1 and 2.
A:	affinity.
$\tilde{\mathrm{A}}$:	electrochemical affinity.
A_M:	molar area.
A_m:	molecular area.
a:	cohesion pressure of a gas or radius of the unit cell of a liquid.
A, B, …:	components of a mixture.
a^{mix} and b^{mix}:	mixing terms of the constants in a state equation.
B'_i:	i^{th} coefficient of the virial in the pressure expansion.
B_i:	i^{th} coefficient of the virial.
b:	covolume of a gas or cosurface of an adsorbed gas.
C:	concentration or concentration in a potential-pH plot.
C_P^{xs} :	molar heat capacity of excess at constant pressure.

C_i: molar concentration (or molarity) of component i.

$C_\pm$: mean concentration of ions in an ionic solution.

$C_{V(el)}$: contribution of free electrons in a metal to the molar heat capacity.

$C_{v(r)}$: contribution of rotational motions to the heat capacity at constant volume.

$C_{v(t)}$: contribution of translational motions to the heat capacity at constant volume.

$C_{v(v)}$: contribution of vibrational motions to the heat capacity at constant volume.

C_V, C_P: heat capacity at constant volume and constant pressure, respectively.

c: capacity of a capacitor or number of independent components.

D : dielectric constant of the medium or diameter of protection or contact of a molecule.

$D(T/\Theta_D)$: Debye's function.

d: distance between two liquid molecules.

d_eS: entropy exchange with the outside environment.

d_i: degree of oxidation i of an element A.

d_iS: internal entropy production.

$d\omega$: elementary volume.

E: energy of the system.

E: Young's modulus.

$E(T/\Theta_E)$: Einstein's function.

E_0: internal energy associated with a reaction at a temperature of 0K.

E^0: standard electrical potential or standard electromotive force (emf) of an electrochemical cell.

E_{abs}: reversible emf of an electrochemical cell.

Eb: balance equation.

$\langle E \rangle$: mean total energy of an element in the canonical ensemble.

E_C: total energy of the canonical ensemble.

E_I: potential energy due to interactions.

E_j: energy of an element j of the canonical ensemble.

E_{kin}: molar kinetic energy of electrons in a metal.

E_p: set of variables with p intensive variables chosen to define a system.

e: relative emf of an electrode.

e^0: standard emf of an electrode.

e_0: equi-activity- or equiconcentration emf of an electrode.

e_{abs}: absolute emf of an electrode.

F: Helmholtz energy.

F_m^{mix}: molar excess Helmholtz energy.

$\overline{F}_i^{xs}$: partial molar excess Helmholtz energy of the component i.

$\overline{F}_i^{mix}$: partial molar mixing Helmholtz energy of the component i.

$\overline{F}_i$: free energy, partial molar Helmholtz energy of the component i.

F_{el}: contribution of free electrons to the molar Helmholtz energy.

$\tilde{F}$: electrochemical Helmholtz energy.

F_m: molar Helmholtz energy.

$\mathcal{F}$: faraday.

f_{het}: heterogeneous wetting function.

f_i: fugacity of the component i in a gaseous mixture.

f_i^0: molar Helmholtz energy of pure component i.

f^{θ} or f_i^0: fugacity of a pure gas i.

G_m^{xs} :	excess Gibbs energy.
$\tilde{G}_\sigma$:	electrocapillary Gibbs energy.
$\tilde{G}$:	electrochemical Gibbs energy.
$\overline{G}_i^{xs}$:	partial excess molar Gibbs energy of component *i*.
G, $\overline{G}_i$, $[G]$:	free enthalpy, partial molar free enthalpy of *i*, generalized free enthalpy.
G_m :	molar Gibbs energy.
G_m^{mix} :	molar Gibbs energy of mixing.
g:	osmotic coefficient *or* acceleration due to gravity or degeneration coefficient or multiplicity or statistical mass.
g_i^0 :	molar Gibbs energy of pure component *i*.
g_a:	statistical weight of fundamental electron level of nucleus a.
g_i:	coefficient of multiplicity of state *i*.
$g_{(e)}$:	statistical weight of electron levels.
$g(\mathrm{r})$:	radial distribution function.
$g(v_x)$:	distribution of velocity components along *Ox* axis.
g^*:	molar Gibbs energy of gas *i* at pressure of 1 atmosphere in a mixture.
H_T^0 :	standard molar enthalpy of formation at temperature *T*.
H , $\overline{H}_i$:	enthalpy, partial molar enthalpy of *i*.
H:	Hamiltonian.
$H_{i,}$:	integral of resonance between two neighboring identical atoms.
$H_{i,i}$:	Coulombian integral between two neighboring identical atoms.
$\mathcal{H}$:	magnetic field.

$\tilde{H}$: electrochemical enthalpy.

H_m^{xs} : molar excess enthalpy.

H_m^{mix} : molar mixing enthalpy.

$\overline{H}_i^{xs}$: partial excess molar enthalpy of component *i*.

$\overline{H}_i^{mix}$: partial molar mixing enthalpy of component *i*.

h_t : spreading coefficient.

h: stoichiometric coefficient of protons in an electrochemical reaction.

h: Planck's constant.

h_i^0 : molar enthalpy of pure component *i*.

h_{sp}: Harkins spreading coefficient of a liquid on another.

I: ionic strength of a solution of ions.

I_m : ionic strength in relation to molality values.

I, I_1, I_2, I_3: moments of inertia.

I_t: integral of configuration of the canonical distribution function of translation.

i: Van't Hoff factor.

$\overline{J_i}$: partial molar value of *J* relative to component *i*.

J_i^{mix} : mixing value of *J* relative to component *i*.

$\overline{J_i^{mix}}$: partial molar mixing value of *J* relative to component *i*.

J_i^* : value of *J* relative to component *i* in a perfect solution.

$\overline{J_i^*}$: partial molar value of *J* relative to component *i* in a perfect solution.

j_i^0 :	value of J for the pure component i in the same state of segregation.
j:	rotational quantum number.
$K_{i,j}(\mathrm{E}_p)$:	thermodynamic coefficient associated with the set of variables Ep. X_j is its definition variable and Y_i its definition function.
$K_i^{(\alpha\beta)}$:	constant of change of equilibrium for phase transition Tr for component i.
K_{ij} :	weighting factor of local composition.
K_{ads}:	equilibrium adsorption constant.
K_{AX}:	solubility product of solid AX.
$K_i^{(\alpha\beta)}$:	coefficient of sharing of compound i between the two phases α and β.
K_d:	dissociation constant.
K_{fe}:	adsorption equilibrium function.
$K_r^{(c)}$:	equilibrium constant relative to concentrations.
$K_r^{(f)}$:	equilibrium constant relative to fugacity values.
$K_r^{(P)}$:	equilibrium constant relative to partial pressure values.
K_r:	equilibrium constant.
K_s:	solubility product.
k:	wavenumber.
$\mathrm{k_B}$:	Boltzmann's constant.
L_t:	latent heat accompanying the transformation t.
l_c:	capillary length.
M:	molar mass.
M:	magnetic moment or Madelung constant.

m_s : mass of solute s in grams per kilo of solvent.

m: total mass.

m_i: mass of component i.

N: number of components of a solution or a mixture of gases or involved in a reaction or number of molecules of a collection.

N_a: Avogadro's number.

N_A: number of molecules of component A.

N_C: number of elements in the canonical collection.

N_c: total number of cells of a liquid.

n_i: number of objects i in the system with energy ε_i or number of moles of component i.

n: translational quantum nimber or total number of moles in a solution or a mixture.

$n^{(\alpha)}$: total number of moles in a phase α.

$\langle n \rangle$: mean number of neighboring vacancies of a molecule in a liquid.

N_L: total number of vacancies in a liquid.

P_c^{mix} : critical pressure of the mixture.

P: pressure of a gas.

P_i^{subl} : sublimating vapor pressure of component i.

P_i^{vap} P_i^0 : saturating vapor pressure of component i.

P_r^{mix} : relative pressure of the mixture.

P_c: critical pressure.

P_i: partial pressure of component i.

P_j: proportion of number of elements in a state j.

p: number of external physical variables *or* spreading parameter.

p_F: Fermi pulse.

Q:	heat involved.
Q_a:	reaction quotient in terms of activity.
Q_P:	heat of transformation at constant pressure; quotient of reaction in terms of partial pressures.
Q_r:	reaction quotient of transformation r.
Q_V:	transformation heat at constant volume.
q_ϕ:	equilibrium heat of adsorption.
q_d:	differential heat of adsorption.
q_i:	volumetric fraction parameter.
q_{isost}:	isosteric heat of adsorption.
$\Re$:	reaction rate
R:	perfect gas constant.
R:	mean curvature radius of a surface or rate of reflux of distillation.
r_A :	radius of the ionic atmosphere.
r_0:	minimum distance of energy between two molecules.
r_c:	radius of a cylindrical tube.
r_i:	volumetric fraction parameter.
r_K:	Kelvin radius.
S_m^{mix} :	molar mixing entropy.
$\overline{S}_i^{xs}$:	partial excess molar entropy of component i.
$\overline{S}_i^{mix}$:	partial mixing molar entropy of component i.
S:	oversaturation of a solution.
$\overline{S}_i$:	entropy or partial molar entropy of i.
$\tilde{S}$:	electrochemical entropy.
S_m^{xs}	excess molar entropy.

s: parameter of order of an alloy.

s_i^0: molar entropy of pure component i.

T: temperature.

T_c^{mix}: critical temperature of the mixture.

T^*: second-order transition temperature.

T_r^{mix}: relative temperature of the mixture.

$T_{(Az)}$: boiling point of azeotropic solution.

T_c: critical temperature.

T_F: Fermi temperature.

$T_{i(Eb)}$: boiling point of pure i.

$T_{i(F)}$: melting point of pure i.

T_s: sublimation temperature.

T_v: vaporization temperature.

U_m^{xs}: excess molar internal energy.

U_m^{mix}: mixing molar internal energy.

$\overline{U}_i^{xs}$: excess partial molar internal energy of component i.

$\overline{U}_i^{mix}$: partial mixing molar internal energy of component i.

U, $\overline{U}_i$: internal energy, partial molar internal energy of i.

U_{el}: contribution of free electrons to the molar internal energy.

$\tilde{U}$: internal electrochemical energy.

U_m: molar internal energy.

U_R: crosslink internal energy.

u_+, u_-: ionic mobilities of the cation and anion.

u_i^0 : molar internal energy of pure component *i*.

$V, \overline{V}_i$: volume, partial molar volume of *i*.

V_c: critical volume.

V_G: Gibbs variance.

V_m: molar volume.

v_c: volume of the unitary element of a liquid.

v_D: Duhem variance.

v_f: free volume per molecule.

v_∞: volume of influence around a molecule.

v_i^0 : molar volume of pure component *i*.

v: quantum vibration number.

v_m: molecular volume.

v_M: molar volume of solid at melting point.

v_{mono}: volume of monolayer of adsorbed gas.

v_{xi}: component along *Ox* axis of the velocity of a particle *i*.

W_{12}: energy per square meter of interaction between the surfaces of phases 1 and 2.

w_i: mass fraction of the component *i*.

w_{ij}: energy of exchange between atoms *i* and *j*.

$x_k^{(\alpha)}$: mole fraction of component *k* in phase α.

x, *y*, *z*: coordinates of a point in space.

x_i: molar fraction of the component *i* in a solution.

$\langle y \rangle$: mean value of *y*.

Y_i and X_i: conjugal intensive and extensive values.

$y_{i,j}$: Mayer function.

y_i: molar fraction of component *i* in a gaseous phase.

Z: compressibility coefficient.

Z_i: compressibility coefficient of gas *i*.

Z^{mix}: compressibility coefficient of the mixture of gases.

Z_{AB}: molecular partition function of interaction between molecules.

Z_C: canonical partition function.

$Z_{C(\mathrm{A})}$: canonical partition function of component A.

$Z_{C(I)}$: canonical partition function of interaction.

$Z_{C(t)}$: canonical partition function of translation.

z: molecular partition function, altitude of a point or coordination index, number of nearest neighbors.

z_e: electron molecular partition function or electrovalence of ion *i*.

z_i: number of molecules that are near neighbors of a molecule *i*.

z_{int}: contribution of internal motions to the molecular partition function.

z_n: molecular partition function of nuclei.

z_{pf}: molecular partition function of a perfect gas.

z_r: rotational molecular partition function.

z_t: translational molecular partition function.

$z_{t(pf)}$: translational molecular partition function of a perfect gas.

z_v: vibrational molecular partition function.

α: coefficient of dissociation of a weak electrolyte or linear dilation coefficient at pressure *P* or relative volatility or Lagrange multiplier relating to the number of objects of a collection or polarizability of a molecule.

α_a: apparent dissociation coefficient of a weak electrolyte.

β: Lagrange multiplier relating to the energy of the objects in a collection or volumetric dilation coefficient at pressure *P*.

$\Gamma(\mathrm{E}_P)$: characteristic function with the set E_P as canonical variables.

Γ_k : coefficient of activity of a group.

Γ: characteristic function.

Γ_i: surface excess or surface concentration of component *i*.

$\Gamma_{i,j}$: excess surface or surface concentration of component *i* in relation to *j*.

γ: coefficient of activity of the component *i* irrespective of the reference state or Grüneisen parameter or structure coefficient whose value is $\sqrt{2}$ for cubic crystal lattices with centered faces.

γ_0: activity coefficient of a solvent.

γ_i: activity coefficient of the species *i* or Grüneisen factor of phonon *i*.

$\gamma_i^{(I)}$: activity coefficient of component *i*, pure-substance reference.

$\gamma_i^{(II)}$: activity coefficient of component *i*, infinitely dilute solution reference.

$\gamma_i^{(III)}$: activity coefficient of component *i*, molar solution reference.

$\gamma_\pm$: mean activity coefficient of ions in an ionic solution.

γ_s: activity coefficient of a solute.

$\Delta\sigma$: spreading of a liquid.

$\Delta_r(A^0{}_T)$: standard value at temperature *T* of *A* associated with the transformation *r*.

$\Delta_r A$: value de *A* associated with the transformation *r*.

$\delta_{i,j}$: Kronecker delta.

δ: coefficient of pressure increase at volume *V*.

$\varepsilon_{\mathrm{A(A)}}$: network energy of an atom of A in network A.

ε_i^j : Wagner interaction coefficient.

ε : electrical permittivity of the medium.

ε_0 : electrical permittivity of a vacuum.

ε_{attr}: energy of attraction between molecules.

ε_c: kinetic energy of a molecule.

$\varepsilon_{\text{C-H}}$: energy of the C-H bond.

ε_d: energy from the dispersion effect between molecules.

ε_F: Fermi energy.

$\varepsilon_{i(e)}$: electronic energy of a molecule *i*.

$\varepsilon_{i(I)}$: interactional energy of a molecule *i*.

$\varepsilon_{i(n)}$: nuclear energy of a molecule *i*.

$\varepsilon_{i(r)}$: rotational energy of a molecule *i*.

$\varepsilon_{i(t)}$: translational energy of a molecule *i*.

$\varepsilon_{i(v)}$: vibrational energy of a molecule *i*.

$\varepsilon_{i,j}$: energy of interaction between two molecules *i* and *j* or pair energy between atoms *i* and *j*.

ε_{mm}: switch.

ε_o: energy due to the effect of orientation between molecules.

ε_p: potential energy of a molecule.

ε_{rep}: repulsion energy between molecules.

η: viscosity.

η_{ij} : Warren and Cowley's order parameter.

Θ_{D}: Debye's vibration temperature.

Θ_{E}: Einstein's vibration temperature.

Θ_{r}: characteristic rotation temperature.

θ: overlap fraction.

θ_i: surface fraction of a component.

λ:	linear dilation coefficient.
λ_{0+}, λ_{0-}:	equivalent ionic conductivities of the cation and anion.
λ_A :	absolute activity of component A.
λ_i :	lateral chemical potential of component *i*.
Λ:	equivalent conductivity of an electrolyte or thermal wavelength of a molecule.
Λ_0:	maximum equivalent conductivity of an electrolyte.
μ_i, $[\mu_i]$, $\overline{\mu}_i$:	chemical potential of the component *i*, dipolar electrical moment of molecule *i*, generalized chemical potential.
$\mu_i^{(L)}, \mu_i^{(G)}$:	chemical potential of component *i* in liquid/gaseous state, respectively.
$\tilde{\mu}$:	electrochemical potential.
ν:	vibration frequency.
$\nu_{k(\rho)}$:	algebraic stoichiometric number of component A_k in reaction ρ.
ν_D:	Debye's maximum frequency.
ν_e:	stoichiometric coefficient of electrons in an electrochemical reaction.
ξ:	reaction extent.
Π_d:	disjunction pressure.
ρ:	density of molecules in a spherical crown of radius *r* or volumetric density of electrical charges or density.
$\rho(r)$:	density of molecules in an enclosure.
σ:	surface energy or symmetry number.
σ_e:	surface density of electrical charges.
σ^*:	surface tension.
τ_+, τ_-:	cationic and anionic transport numbers.
Φ :	practical osmotic coefficient; expansion pressure.

Φ_i : coefficient of fugacity of component i in a gaseous mixture.

ϕ: coefficient of conductivity of a strong electrolyte or number of Phases.

ϕ_i : coefficient of fugacity of gas i in a mixture or volume fraction of a component.

ϕ^0 or ϕ_i^0 : coefficient of fugacity of a pure gas.

χ_i : calorimetric coefficient relative to the variable x_i.

χ: electrical conductivity.

χ_T: coefficient of compressibility at temperature T.

Ψ_i : electrostatic potential of ionic atmosphere.

$\Psi(r)$: electrostatic potential.

Ψ_{km} : energy term between two groups.

Ψ: wavefunction.

Ω_{BE}: number of complexions in Bose-Einstein statistics.

Ω_C: number of complexions in Fermi-Dirac statistics.

Ω: number of complexions.

ω_i: set of position coordinates of molecule i.

ω_x: rotational velocity component in direction Ox.

1

Pure Crystalline Solids

Crystalline solids are characterized by the regular and periodic spatial arrangement of entities at the nodes of a lattice. The nature of the entities thus arranged defines the nature of the solid. There are four distinct classes:

– atomic solids, comprising a lattice of atoms, such as solid argon, for example;

– molecular solids, where the entities arranged at the nodes of the lattice are molecules, as is the case in solid benzene;

– ionic crystals. In this case, the entities are ions, and they are arranged into two sublattices: one of cations and the other of anions. The proportions of sites occupied by these two sublattices are obviously such that the whole solid is electrically neutral, overall. The ions thus arranged could either be simple ions, as is the case with sodium chloride, or complex ions such as in ammonium carbonate;

– metals, in which ions are arranged at the nodes of the lattice. To ensure electrical neutrality, more-or-less mobile electrons are distributed around these ions.

1.1. Characteristic values of a solid

Solids are incompressible, which means that their derivative $(\partial V / \partial P)_T$ is practically zero, so they do not have an equation of state such as $F(P, V, T) = 0$.

However, solids do experience changes in volume, under the influence of temperature, which is characterized by its cubic expansion coefficient or its linear expansion coefficient.

Similarly, when heat is applied to it, solids heat up. The extent of that rise in temperature is characterized by the solid's specific heat capacity.

When a solid is subjected to a stress (or load) – i.e. a certain amount of force per unit surface area or a moment per unit length, such as axial traction (Figure 1.1(a)) or axial compression (Figure 1.1(b)), for example – it normally deforms. This deformation is also known as "strain".

Other values pertaining to the dielectric and magnetic properties are also available.

In the next section, we will examine the first three effects, starting with the effect of a stress. Then we will develop models of solids which we can go on to use in dealing with the questions of specific heat capacities and thermal expansion.

1.2. Effect of stress and Young's modulus

When a solid is subjected to a stress, it generally experiences a strain which, if pursued, could cause the material to fracture. The applied stress is measured in newtons per square meter.

Figure 1.1. *Cylindrical test tube a) under traction; b) under compression*

Take the example of traction applied to a cylinder whose initial length is l_0 and cross-section area is s (Figure 1.1(a)). If we begin with zero stress, and very gradually increase the stress (i.e. the traction force) at a constant temperature, the relative strain $\Delta l/l_0$ increases, obeying a law which is often identical to its tangent to the origin – that is, a practically linear law (the part "OA" of the curve shown in Figure 1.2) – which is known as Hooke's law, and is written as follows for a given temperature:

$$\frac{F}{s} = \mathrm{E}\frac{\Delta l}{l_0} \qquad [1.1]$$

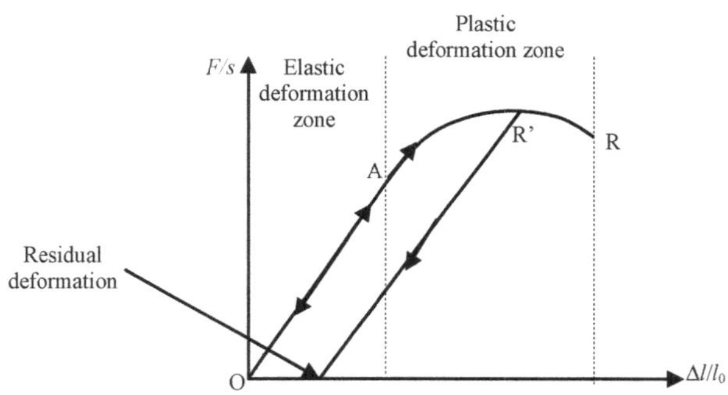

Figure 1.2. *Strain/stress curve under traction*

A solid, therefore, is characterized by its modulus of longitudinal elasticity, known as Young's modulus, defined by:

$$\mathrm{E} = \frac{l_0 s}{\dfrac{1}{\left(\dfrac{\partial l}{\partial F}\right)}} \qquad [1.2]$$

If we start at O and arrive at a point between O and A (a point marking the start of a lesser increase of the curve), and we slowly decrease the stress, the strain decreases in accordance with the same Hooke curve. We then say that we are in the domain of elastic deformation. When the stress returns to zero, the sample returns to its initial length. The load at point A is called the elastic limit.

After point A in Figure 1.2, the curve changes direction, and Hooke's law is no longer obeyed. If we stop at a point R' between A and R, and decrease the stress, the curve then follows a straight line which is practically parallel to Hooke's straight line. The result of this is that when the stress returns to zero, there is a remaining deformation which is known as the residual deformation. We say that between points A and R, the solid undergoes a plastic deformation.

When we reach a point R, the test-tube breaks. The corresponding stress is called the fracture limit.

The different characteristics of the curves in Figure 1.2 vary greatly from one material to another. Hence, for example, the fracture load is much greater for steel than for concrete, which has pitiful traction resistance – this is why steel rods are used in reinforced concrete.

NOTE.– We have pointed out that the stress variations must, during increase and decrease, take place very slowly, because strictly speaking, the Young's modulus depends not only on the temperature but also on the rate at which the load changes.

The Young's modulus obviously depends on the temperature, but these data can sometimes be replaced by a state equation in the form F(F,l,T) = 0, such as that given for rubber, which links the temperature T, the length l of a cylinder of rubber with section s, to the force of traction F exerted upon it in the elastic domain by way of two constants, B and its cubic expansion coefficient β which is of the form:

$$F = BT\left\{\frac{l}{l_0} - \left[1+\beta(T-T_0)\right]\left(\frac{l_0}{l}\right)^2\right\} \quad [1.3]$$

We can obtain very similar results by imposing other types of stresses such as compression or shear. Of course, the corresponding moduli have completely different values from those encountered in the case of traction.

1.3. Microscopic description of crystalline solids

From a microscopic point of view, we find the four classes of solids introduced at the start of this chapter.

Based on each of these models, we will establish canonical partition functions which define the microscopic system, followed by the Helmholtz energy, which links the microscopic point of view to the macroscopic system, and finally the internal energy, which will be useful for us later on to establish the specific heat capacities.

1.4. Partition function of vibration of a solid

Irrespective of the nature of the solid, there are always species (atoms, ions and molecules) placed at the nodes of the lattice, and are animated with a motion of vibration around their equilibrium positions. Thus, the partition function will include a contribution due to these vibrations. On the statistical level, the entities in question (atoms, ions or molecules) are considered to be localized particles and, in general, to describe the vibrations of the solid, it is sufficient to place ourselves in the context of the conventional limit case of statistics. It results from this that the contribution Z_C of the vibrations to the canonical partition function can be calculated on the basis of the atomic partition functions z by the relation:

$$Z_C = z^N \qquad [1.4]$$

We will calculate this contribution by first considering a solid as a macromolecule with N vibrating entities, having $3N$ independent vibrational degrees of freedom: phonons. Two models can be used to express a contribution of these vibrations to the canonical partition function: Einstein's model and Debye's, which is more general. Let us look at each of these two models in turn.

1.4.1. *Einstein's single-frequency model*

In this model, the crystalline solid is a system of $3N$ oscillators, in which all have the same fundamental frequency ν_E. The corresponding characteristic temperature of vibration, defined on the basis of the relation:

$$\Theta_E = \frac{h\nu}{k_B} \qquad [1.5]$$

is called the *Einstein temperature* Θ_E. By applying equation [1.4] and the expression of the vibrational partition function, written as:

$$z_v = \frac{\exp\left[-\frac{\Theta_E}{2T}\right]}{1-\exp\left[-\frac{\Theta_E}{T}\right]} \qquad [1.6]$$

we can write the canonical partition function of vibration in the form:

$$Z_{C(v)} = (z_v)^{3N} = \exp\left(-\frac{3N\Theta_E}{2T}\right)\left[1-\exp\left(-\frac{\Theta_E}{T}\right)\right]^{-3N} \qquad [1.7]$$

or if we switch to the logarithm, the previous equation becomes:

$$\ln Z_{C(v)} = -\frac{3N\Theta_E}{2T} - 3N\ln\left[1-\exp\left(-\frac{\Theta_E}{T}\right)\right] \qquad [1.8]$$

1.4.2. *Debye's frequency distribution model*

In the case of Debye's model, the $3N$ oscillators no longer have the same fundamental frequency. By applying the equation:

$$z_v \cong \prod_E \frac{\exp\left(-\frac{\Theta_E}{2T}\right)}{1-\exp\left(-\frac{\Theta_E}{T}\right)} \qquad [1.9]$$

we are able to write the following for the canonical partition function of vibration:

$$Z_{C(v)} = \prod_{i=1}^{3N} \frac{\exp\left(-\frac{h\nu_i}{2k_B T}\right)}{1-\exp\left(-\frac{h\nu_i}{k_B T}\right)} \qquad [1.10]$$

However, if we reformulate in logarithmic terms, we find:

$$\ln Z_{C(\nu)} = \frac{\sum_{i=1}^{3N} \mathrm{h}\nu_i}{2\mathrm{k_B}T} - \sum_{i=1}^{3N} \ln\left[1 - \exp\left(-\frac{\mathrm{h}\nu_i}{\mathrm{k_B}T}\right)\right] \quad [1.11]$$

Debye supposes that the frequencies are sufficiently similar so that the distribution can be supposed to be continuous, which enables us to replace the sum of equation [1.11] by an integral, which is written as follows for a frequency distribution $g(\nu)$:

$$\ln Z_{C(\nu)} = -\frac{\int_0^{\nu_D} \mathrm{h}\nu g(\nu)\mathrm{d}\nu}{2\mathrm{k_B}T} - \int_0^{\nu_D} \ln\left[1 - \exp\left(-\frac{\mathrm{h}\nu}{\mathrm{k_B}T}\right)\right] g(\nu)\mathrm{d}\nu \quad [1.12]$$

Debye also supposes that this frequency distribution is of the same form as the elastic frequency distribution of the solid, supposed to be a continuum. Those frequencies themselves are linked to the propagation of sound in that solid. These frequencies range from the value 0 to a maximum frequency ν_D defined by equation [1.13], where c is the celerity of sound in the solid.

$$\nu_D = c\left(\frac{3N}{4\pi V}\right)^{1/2} \quad [1.13]$$

The corresponding frequency temperature – defined, as for Einstein's temperature, using relation [1.5] – which is called the *Debye temperature* Θ_D, is thus defined by:

$$\Theta_D = \frac{\mathrm{h}\nu_D}{\mathrm{k_B}} \quad [1.14]$$

Table 1.1 shows a selection of values of the Debye temperature which can, as we can see, be very different depending on the solid.

Solid	Θ_D(K)
Pb	88
Ag	220
Al	396
KCl	227
C (diamond)	2067

Table 1.1. *The Debye temperature of various solids (data taken from [INF 06])*

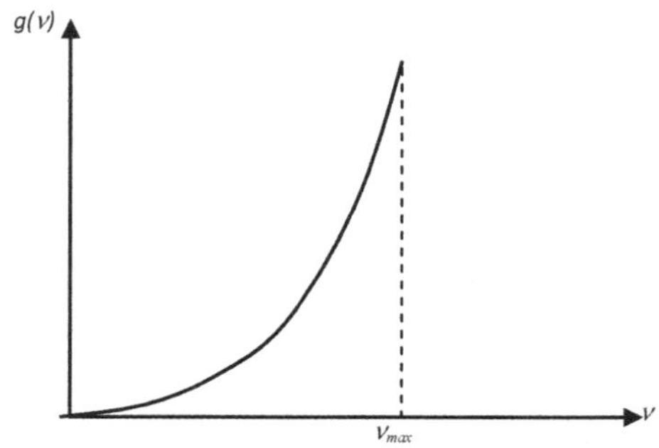

Figure 1.3. *Frequency distribution according to Debye*

The frequency distribution $g(\nu)$ is of the form (see Figure 1.3):

$$\begin{cases} \text{For } \nu \leq \nu_D : g(\nu) = \dfrac{9N\nu^2}{\nu_D^3} \\ \text{For } \nu > \nu_D : g(\nu) = 0 \end{cases} \quad [1.15]$$

obviously, with the standardization condition:

$$\int_0^{\nu_D} g(\nu)\mathrm{d}\nu = 3N \quad [1.16]$$

Relation [1.12] then takes the form:

$$\ln Z_{C(\nu)} = -\frac{9N\Theta_D}{8T} - \frac{9N}{\nu_D^3}\int_0^{\nu_D} \nu^2 \ln\left[1 - \exp\left(-\frac{\mathrm{h}\nu}{\mathrm{k_B}T}\right)\right]\mathrm{d}\nu \quad [1.17]$$

If we set:

$$x = \frac{\mathrm{h}\nu}{\mathrm{k_B}T} \qquad [1.18]$$

relation [1.17] can also be written:

$$\ln Z_{C(\nu)} = -\frac{9N\Theta_D}{8T} - \frac{9NT^3}{\Theta_D^3}\int_0^{\Theta_D/T} x^2 \ln\left[1-\exp(-x)\right]\mathrm{d}x \qquad [1.19]$$

The integral appearing in this formula cannot be expressed by a simple algebraic sum. However, if we integrate by parts, we find:

$$\ln Z_{C(\nu)} = -\frac{9N\Theta_D}{8T} - 3N\ln\left[1-\exp\left(-\frac{\Theta_D}{T}\right)\right] + \frac{3NT^3}{\Theta_D^3}\int_0^{\Theta_D/T} \frac{x^3\exp(-x)}{1-\exp(-x)}\mathrm{d}x \qquad [1.20]$$

1.4.3. *Models with more complex frequency distributions*

Debye's model, which allows for a frequency distribution given by relation [1.9], has only yielded correct values of the specific heat capacity at constant volume (see section 1.8) for fairly low temperatures. Other authors have improved the model by modifying that frequency distribution. For example, Born and Karman took a new approach to the establishment of the frequency distribution, this time supposing that the solid was no longer a continuum, but instead was represented by a periodic lattice of particles, which led them to the distribution function as shown in Figure 1.4(a). The distribution function reaches its peak very near to the limit frequency.

Blackman, for his part, determined the vibration spectrum for simple cubic lattices. The frequency distribution which it achieves exhibits two maximum points (see Figure 1.4(b)). One point of these is always situated in the vicinity of the limit frequency; the other point, which is less clear, is at a lower frequency.

In fact in this types of models, with a frequency distribution modified compared to Debye's, we can keep the developments obtained using Debye's model, but as if the Debye temperature varied with temperature.

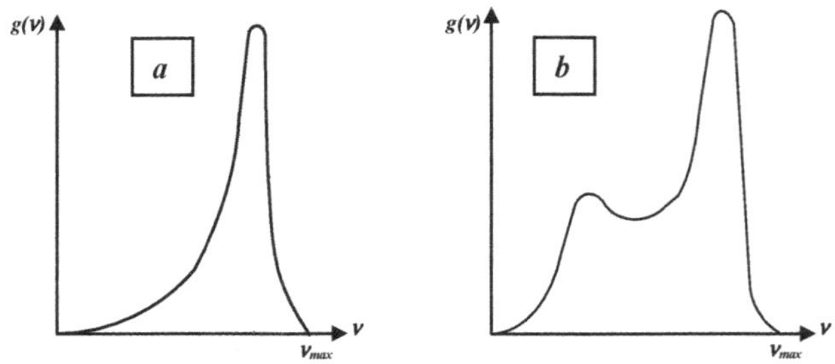

Figure 1.4. *Frequency distributions: a) Born and Karman; b) Blackman*

Variations in the Debye temperature with the temperature have, indeed, been observed by Mott and Jones.

1.5. Description of atomic solids

In an atomic solid, each node of the lattice is occupied by an atom, e.g. in the case of a solid rare gas, but also chlorine, fluorine, etc. For a pure solid, there is only one sort of atom.

1.5.1. *Canonical partition function of an atomic solid*

Besides the vibrating motion of the atoms which we have discussed above, the solid is home to electron motion and nuclear spin, each of which makes a contribution to the canonical partition function.

With regard to the motion of the electrons, we choose as energy origin the fundamental level in the atom, and these electrons are not excited, so that the contribution to the atomic partition function of electrons is reduced to its statistical weight, according to the equation $z = g_0$, and therefore the electron contribution to the canonical partition function will be:

$$\ln Z_{C(e)} = N \ln g_{0(e)} \qquad [1.21]$$

For the contribution of nucleus spin, we see the same situation as for the electrons, so their contribution will be:

$$\ln Z_{C(n)} = N \ln g_{0(n)} \qquad [1.22]$$

The overall canonical partition function for the solid will be:

$$\ln Z_C = \ln Z_{C(v)} + \ln Z_{C(e)} + \ln Z_{C(n)} \qquad [1.23]$$

However, it is helpful, in studying solids, to choose as the energy origin the atoms which are infinitely far apart, instead of the fundamental level of vibration. Therefore, for each atom, we need to involve the crosslink energy u_R, which is the energy needed to send that atom to infinity. This introduces a new term, known as crosslinking – a relative distribution function of the atoms – into the canonical partition function, and gives us a term in the following form:

$$\ln Z_{C(r)} = \frac{N u_R}{k_B T} = \frac{U_R}{k_B T} \qquad [1.24]$$

In light of relations [1.21], [1.22] and [1.24], expression [1.23] becomes:

$$\ln Z_{C(v)} = \frac{U_R}{k_B T} - \frac{9N\Theta_D}{8T} - 3N \ln\left[1 - \exp\left(-\frac{\Theta_D}{T}\right)\right] + \frac{3NT^3}{\Theta_D^3} \int_0^{\Theta_D/T} \frac{x^3 \exp(-x)}{1 - \exp(-x)} \mathrm{d}x + N \ln g_e g_n \qquad [1.25]$$

1.5.2. *Helmholtz energy and internal energy of an atomic solid*

The Helmholtz energy F of a solid, comprising N atoms, can be calculated using the relation that links Helmholtz energy to the canonical partition function:

$$F = -k_B T \ln Z_C \qquad [1.26]$$

In view of the canonical partition function, given by relation [1.25], the molar Helmholtz energy F_m is obtained by applying the two relations $N = \mathrm{N_a}$ and $\mathrm{k_B N_a} = \mathrm{R}$ in expression [1.26]. From this, we can easily deduce the molar internal energy using the relation:

$$U_m - U_0 = -\frac{1}{\mathrm{R}T^2} \frac{\mathrm{d}\left(\frac{F_m}{T}\right)}{\mathrm{d}T} \qquad [1.27]$$

In fact, that internal energy is the sum of the four contributions: crosslink (independent of temperature), vibration (a function of the temperature), electron and nuclear activity (practically independent of temperature), which is written as:

$$U_m - U_0 = U_{m(R)} + U_{m(v)}(T) + U_{m(e)} + U_{m(n)} \qquad [1.28]$$

For the contribution of vibration, we can either use Einstein's model, which yields:

$$U_{m(v)} = \frac{3\mathrm{R}\Theta_E}{\left[\exp\left(-\frac{\Theta_E}{T}\right) - 1\right]} \qquad [1.29]$$

or Debye's model, which gives us:

$$U_{m(v)} = \frac{9\mathrm{R}T^4}{\Theta_D^3} \int_0^{\Theta_d/T} \frac{x^3}{\left[\exp(x) - 1\right]} \mathrm{d}x \qquad [1.30]$$

We cannot analytically calculate the integral appearing in relation [1.30], but we can give two approximate values for it depending on whether the temperature is high or low in relation to the Debye temperature.

For the high temperatures ($T >> \Theta_D$), we develop the function to be integrated into a Maclaurin series and integrate, we find:

$$U_{m(v)} = -\frac{9\mathrm{R}\Theta_D}{8} + 3\mathrm{R}T\left[1 + \frac{1}{20}\left(\frac{\Theta_D}{T}\right)^2 + \mathrm{O}\left(\frac{\Theta_D}{T}\right)^4\right] \qquad [1.31]$$

The term $\mathrm{O}\left(\frac{\Theta_D}{T}\right)^4$ indicates the rest of the series development.

For low temperatures ($T \ll \Theta_D$), we replace the integral with the difference of two integrals – the first between 0 and infinity and the second between Θ_D / T and infinity. We can show that we obtain:

$$U_{m(v)} = 3\mathrm{R}T\left[\frac{\pi^4 T^3}{5\Theta_D^3} - 3\exp\left(-\frac{\Theta_D}{T}\right) + \mathrm{O}\left(\frac{\Theta_D}{T}\exp\left(-\frac{\Theta_D}{T}\right)\right)\right] \quad [1.32]$$

1.6. Description of molecular solids

If, instead of one atom per crystalline site, a solid contains a group of atoms (ions or molecules) at each site, if the crystal contains N molecules, the lattice will always have $3N$ degrees of vibrational freedom, but there will still be $3(s^{-1})$ internal degrees of freedom per molecule if the molecule contains s atoms.

1.6.1. *Partition function of molecular crystals*

The new "internal" degrees of freedom will generally tend to be vibrational degrees of freedom, and if the corresponding fundamental frequencies are ν_1, ν_2, ..., ν_i, ..., the contribution corresponding to the canonical partition function is:

$$\ln Z_{C(\text{int})} = N\sum_{i=4}^{3s}\ln\left[1-\exp\left(-\frac{\Theta_i}{T}\right)\right] \quad [1.33]$$

In Einstein's model, in view of the new electronic and nuclear contributions, this would give us:

$$\begin{aligned}\ln Z_C = &-\frac{U_0}{\mathrm{k_B}T} - \frac{3N\Theta_E}{2T} - 3N\ln\left[1-\exp\left(-\frac{\Theta_E}{T}\right)\right] \\ &- N\sum_{i=4}^{3s}\ln\left[1-\exp\left(-\frac{\Theta_i}{T}\right)\right] + N\sum_{i=1}^{s}\ln g_{s(e)} + N\sum_{i=1}^{s}\ln g_{s(n)}\end{aligned} \quad [1.34]$$

and in Debye's model, we would have:

$$\ln Z_C = -\frac{9N\Theta_D}{8T} - 3N\ln\left[1-\exp\left(-\frac{\Theta_D}{T}\right)\right]$$
$$+\frac{3NT^3}{\Theta_D^3}\int_0^{\Theta_D/T}\frac{x^3\exp(-x)}{1-\exp(-x)}\mathrm{d}x - N\sum_{i=4}^{3s}\ln\left[1-\exp\left(-\frac{\Theta_i}{T}\right)\right] \quad [1.35]$$
$$+N\sum_{i=1}^{s}\ln g_{i(e)} + N\sum_{i=1}^{s}\ln g_{i(n)}$$

It is possible that two or three of these degrees of freedom internal to the molecule may be better described as rotations than as vibrations. Such would be the case, for example, with the H_2 molecule. For molecules containing only one atom other than hydrogen atoms – e.g. ClH, CH_4, NH_4^+– we obtain better results when we consider that a movement is indeed a high-temperature rotation, but also a low-temperature vibration. There would be a rather sharp transition within a certain temperature range. In the case of a rotation, a rotational partition function term replaces a vibrational term in equation [1.30] and the corresponding terms in relations [1.30] and [1.32].

1.6.2. *Thermodynamic functions of molecular solids*

As is the case with atomic solids, we can calculate the Helmholtz energy F of a solid containing N molecules by applying equation [1.26] using the canonical partition function given by equation [1.34] or [1.35]. The molar Helmholtz energy F_m is deduced in the same manner by applying the two relations N=N_a and $k_B N_a$ = R in the expressions thus obtained. We can easily deduce the molar internal energy from relation [1.26]. In fact, this internal energy is the sum of five terms:

$$U_m - U_0 = U_{m(R)} + U_{m(v)}(T) + U_{m(e)} + U_{m(n)} + U_{m(\text{int})} \quad [1.36]$$

The contributions $U_{m(R)}$ and $U_{m(v)}$ are the same as with an atomic solid. Thus the contribution of the internal movements in the molecule is expressed as:

$$U_{m(\text{int})} = \frac{\mathrm{R}}{T^4} \sum_{i=4}^{s} \frac{\Theta_i \exp\left(-\frac{\Theta_i}{T}\right)}{1 - \exp\left(-\frac{\Theta_i}{T}\right)} \qquad [1.37]$$

The electron and nuclear contributions are also sums on the number of atoms contained in the molecule:

$$U_{m(e)} = \frac{\mathrm{R}}{T^2} \sum_{i=1}^{s} \frac{\mathrm{d} \ln g_{i(e)}}{\mathrm{d}T} = 0 \qquad [1.38]$$

$$U_{m(n)} = \frac{\mathrm{R}}{T^2} \sum_{i=1}^{s} \frac{\mathrm{d} \ln g_{i(n)}}{\mathrm{d}T} = 0 \qquad [1.39]$$

1.7. Description of an ionic solid

An ionic solid is made up of two kinds of ions: anions and cations. These ions are located at the nodes of two interlocking periodic lattices: the anionic sublattice and the cationic sublattice.

There are two distinct types of ionic solid:

– simple ionic solids, where the ions contain only one atom;

– complex ionic solids, containing at least one type of ion. Usually, the anion is a complex ion, which means that it is a molecule made up of several atoms, which carries an electrical charge.

1.7.1. *Crosslink energy of an ionic solid*

We model the energy of an ionic crystal limiting us to the case of a diatomic crystal with the formula A_aB_b linking a cation A with electrovalence z_A and an anion B with electrovalence z_B. We assume that these ions are comparable to spheres on which the charges are uniformly distributed. The overall cohesion of the crystal is provided by the Coulombian field at $1/r^2$ created by each ion, plus a repulsive potential especially sensitive over a short distance, but which cannot be overlooked if we want to give an account of an equilibrium state of the crystal.

1.7.1.1. *Attraction energy*

Let us consider an ion i in the lattice, bearing the charge q_i. At that point, there is an electrostatic potential Φ_i, created by all the other ions, and the energy of interaction between that ion i and the rest of the lattice is:

$$U_i = q_i \Phi_i \qquad [1.40]$$

In order to find the total attraction energy, we only need to add the above expression for all the ions, taking care only to count the interaction between each pair of ions once (which we do by using the coefficient ½), as follows:

$$U_i = \frac{1}{2}\sum_i q_i \Phi_i \qquad [1.41]$$

In a large crystal, it can be assumed that at each node A of the lattice, the potential is identical – i.e. Φ_A – and at each node B, the potential will be Φ_B (so we overlook any possible edge effects). Thus, if we compare the electrostatic energy to a mole, where e is the elementary charge, we find the following for the molar attraction term:

$$U_{attr} = \frac{N_a}{2}\left(az_A e\Phi_A + bz_B e\Phi_B\right) \qquad [1.42]$$

As a unit of length, let us choose a distance d which is characteristic of the lattice. An ion i will be at distance $d(\rho_i)_A$ from an arbitrary origin, chosen in place of an ion A and at distance $d(\rho_i)_B$ from another arbitrary origin, chosen in place of an ion B. The potentials Φ_A and Φ_B at those origins are, respectively:

$$\Phi_A = \frac{1}{\varepsilon_0 d} \sum_{\mathrm{i}} \frac{z_i e}{(p_i)_a} \qquad [1.43]$$

and

$$\Phi_B = \frac{1}{\varepsilon_0 d} \sum_{\mathrm{i}} \frac{z_i e}{(p_i)_a} \qquad [1.44]$$

Because of the principle of electrical neutrality of the crystal, we must have:

$$az_A + bz_B = 0$$

Thus:

$$\frac{z_A}{b} = \left|\frac{z_B}{a}\right| = \varpi \qquad [1.45]$$

The attraction energy per mole becomes:

$$U_{attr} = \frac{\mathrm{N_a e^2}\varpi^2 ab}{2\varepsilon_0 d}\left(\sum_i \frac{\theta_i}{(p_i)} - \sum_i \frac{\theta_i}{(p_B)}\right) \qquad [1.46]$$

θ_i is equal to b if the ion i is of type A, and to $(-a)$ if the ion i is of type B. This energy can be written in the form:

$$U_{attr} = \frac{\mathrm{N_a} e^2 \varpi^2}{\varepsilon_0 d}\mathsf{M} \qquad [1.47]$$

including a value M, known as the Madelung constant, which depends solely on the nature of the lattice and is defined by:

$$\mathsf{M} = \frac{ab}{2}\left(\sum_i \frac{\theta_i}{(p_i)} - \sum_i \frac{\theta_i}{(p_B)}\right) \qquad [1.48]$$

There are various algorithms which can be used to calculate the Madelung constant. Table 1.2 shows the values obtained for cubic lattices.

Type of lattice	Madelung constant
Sodium chloride	1.74756
Cesium chloride	1.76267
Blende	1.63806
Wurtzite	1.641

Table 1.2. *Values of the Madelung constant in different cubic lattices*

1.7.1.2. *Repulsion energy*

The previous expression of electrostatic energy gives an energy that can only grow in absolute value if the crystalline lattice parameter d of the crystalline lattice decreases. No equilibrium state can be achieved. To explain this, we must introduce a repulsion term, whose intensity will decrease as the distance d increases. Born and Mayer put forward an expression for this repulsion energy arising from the repulsion of the electron clouds if they intermingle. This energy has an exponential form that seems consistent with quantum mechanics. For the interaction of an ion pair i and j located a distance $d\rho_{ij}$ apart, we set:

$$U_{rép(i,j)} = mc_{i,j} \exp\left(-\frac{d\rho_{i,j}}{\rho}\right) \qquad [1.49]$$

In this expression, $c_{i,j}$ and ρ are constants. Authors have shown that we can take the value 0.345×10^{-10} m for ρ for all ions, and that the constant $c_{i,j}$ is given by:

$$c_{i,j} = \left(1+\frac{z_i}{n_i}+\frac{z_j}{n_j}\right)\exp\left(\frac{r_i + r_j}{\rho}\right) \qquad [1.50]$$

In this expression, z_i and z_j are the electrovalences of the ions i and j, n_i and n_j are the numbers of electrons on the outermost layer, r_i and r_j are the ionic radii and m is a new constant, which we will discuss later on.

Looking again at the case of a biatomic solid: we still need to consider the repulsion energy associated with an ion A taken as the origin of the system, and with an ion B taken as the origin. However, in the case of repulsion, the potential decreases very quickly as the distance between the ions increases, and the only way to simplify the calculation is to consider the nearest neighbors to a given ion, with opposite signs.

Let us first examine an ion A as the origin. All its near neighbors will be ions B, and there will be a number β of them. The repulsion energy attached to that ion A will be:

$$U_{rep(\mathrm{A,B})} = m\sum_{i=1}^{\beta} c_{\mathrm{A,\,B}} \exp\left(\frac{d\rho_{\mathrm{A,B}_i}}{\rho}\right) \quad [1.51]$$

where:

$$c_{\mathrm{A,B}} = \left(1+\frac{z_{\mathrm{A}}}{n_{\mathrm{A}}}+\frac{z_{\mathrm{B}}}{n_{\mathrm{B}}}\right)\exp\left(\frac{r_{\mathrm{A}}+r_{\mathrm{B}}}{\rho}\right) \quad [1.52]$$

If we now take an ion B as the origin, its nearest neighbors will be α ions A, and its contribution to the repulsion will be:

$$U_{rep(\mathrm{B,A})} = m\sum_{j=1}^{\alpha} c_{\mathrm{B,\,A}} \exp\left(-\frac{d\rho_{\mathrm{B,A}_j}}{\rho}\right) \quad [1.53]$$

where, clearly:

$$c_{\mathrm{A,B}} = c_{\mathrm{B,A}} \quad [1.54]$$

Hence, the repulsion energy per mole can be written as follows, by adding together the two contributions:

$$U_{rep} = \frac{\mathrm{N_a} m}{2} c_{\mathrm{A,B}} \exp\left(-\frac{d}{\rho}\right)\left[\mathrm{a}\sum_{i=1}^{\beta} \exp\rho_{\mathrm{A,B}_i} + \mathrm{b}\sum_{j=1}^{\alpha} \exp\rho_{\mathrm{B,A}_j}\right] \quad [1.55]$$

1.7.1.3. *Crosslink energy*

The crystal's total crosslink energy is given by the sum of the terms of attraction and repulsion, which (in light of expressions [1.47] and [1.55] given above) gives:

$$U_{m(R)} = \frac{\mathrm{N_a} e^2 \varpi^2}{\varepsilon_0 d}\mathfrak{M} + \frac{\mathrm{N_a} m}{2} c_{\mathrm{A,B}} \exp\left(-\frac{d}{\rho}\right)\left[\mathrm{a}\sum_{i=1}^{\beta} \begin{array}{l}\exp\left(\rho_{\mathrm{A,B}_i}\right) + \\ \mathrm{b}\sum_{j=1}^{\alpha} \exp\left(\rho_{\mathrm{B,A}_j}\right)\end{array}\right] \quad [1.56]$$

Thus, in the expression of the crosslink energy, there is a constant m which needs to be determined. We can find its value by looking at the compressibility coefficient. Imagine that we compress a crystal at the temperature of absolute zero. The variation in internal energy will be:

$$\mathrm{d}U = -P\,\mathrm{d}V \quad [1.57]$$

This variation is due only to the change in the crosslink energy. By taking account of the definition of the compressibility coefficient, which is:

$$\chi_T = -\frac{1}{V}\left(\frac{\partial V}{\partial P}\right)_T \quad [1.58]$$

we find:

$$\chi_T V\left(\frac{\mathrm{d}^2 U_{(R)}}{\mathrm{d}V^2}\right) = 1 \quad [1.59]$$

By applying the derivation of a function of function, we can write:

$$\left(\frac{\mathrm{d}^2 U_{(R)}}{\mathrm{d}d^2}\right) = \frac{\mathrm{d}^2 U_{(R)}}{\mathrm{d}V^2}\left(\frac{\mathrm{d}V}{\mathrm{d}d}\right)^2 + \frac{\mathrm{d}U_{(R)}}{\mathrm{d}V}\frac{\mathrm{d}^2 V}{\mathrm{d}d^2} \quad [1.60]$$

For the equilibrium position $d = d_0$, the crosslink energy is minimal, so:

$$\left(\frac{\mathrm{d}U_{(R)}}{\mathrm{d}d}\right)_{d=d_0} = 0 \text{ which gives us } \left(\frac{\mathrm{d}U_{(R)}}{\mathrm{d}V}\right)_{d=d_0} = 0 \quad [1.61]$$

and therefore, equation [1.53] becomes:

$$\left(\frac{\mathrm{d}^2 U_{(R)}}{\mathrm{d}d^2}\right)_{d=d_0} = \left[\frac{\mathrm{d}^2 U_{(R)}}{\mathrm{d}V^2}\left(\frac{\mathrm{d}V}{\mathrm{d}d}\right)^2\right]_{d=d_0} \quad [1.62]$$

The molar volume can always be written in the form:

$$v^0 = \lambda d_0^3 \quad [1.63]$$

with λ being a constant that depends only on the type of lattice. Thus, in view of expression [1.59], relation [1.62] gives us:

$$\left(\frac{\mathrm{d}^2 U_{m(R)}}{\mathrm{d}d^2}\right)_{d=d_0} = \frac{4\lambda^2 d_0^4}{\chi v^0} \quad [1.64]$$

For simplicity's sake, let us write expression [1.56], of the crosslink energy, in the form:

$$U_{m(R)} = -\frac{\mathsf{A}}{d} + \mathsf{B}\frac{m}{\rho}\exp\left(-\frac{d}{\rho}\right) \quad [1.65]$$

where:

$$\mathsf{A} = \frac{\mathrm{N_a e^2}\varpi^2 \mathsf{M}}{\varepsilon_0} \quad [1.66]$$

and:

$$\mathbf{B} = \frac{\mathrm{N_a}}{2} c_{\mathrm{A,B}}\left[\mathrm{a}\sum_{i=1}^{\beta}\exp\left(\rho_{\mathrm{A,B}_i}\right) + \mathrm{b}\sum_{j=1}^{\alpha}\exp\left(\rho_{\mathrm{B,A}_j}\right)\right] \quad [1.67]$$

By twice deriving the function [1.65], we find:

$$\left(\frac{\mathrm{d}^2 U_{m(R)}}{\mathrm{d}d^2}\right)_{d=d_0} = 2\frac{\mathsf{A}}{d_0^3} + \frac{\mathsf{B}m}{\rho^3}\exp\left(-\frac{d_0}{\rho}\right) \qquad [1.68]$$

By equaling the two expressions [1.64] and [1.68] for the second derivative, we obtain:

$$m = \frac{\rho^3}{\mathsf{B}}\left[\frac{4\lambda^2 d_0^4}{\chi v^0} + \frac{2\mathsf{A}}{d_0^3}\right]\exp\left(\frac{d_0}{\rho}\right) \qquad [1.69]$$

In view of equations [1.61] and [1.65], the equilibrium condition is expressed as:

$$\frac{\mathsf{A}}{d_0^2} + \frac{\mathsf{B}m}{\rho^2}\exp\left(-\frac{d_0}{\rho}\right) = 0 \qquad [1.70]$$

Hence, expressions [1.69] and [1.70] constitute a system of two equations where the unknowns are the constants m and d_0, which we can calculate numerically.

1.7.2. *Born/Haber cycle*

Born and Haber envisaged a thermodynamic cycle that can be used to calculate the crosslink energy on the basis of independently-measured data. They begin with the observation that the crosslink energy is the energy that is released if the lattice is formed of gaseous ions which are infinitely immobile in relation to one another. Figure 1.5 shows the cycle for a crystal A_aB_b based on solid metal and a gaseous molecular non-metal, with the symbols having the following meanings:

– *S*: sublimation heat for a gram atom of metal A;

– *D*: dissociation heat for the non-metal B expressed in relation to a gram atom of that substance;

– *I*: ionization energy for a gram atom of metal;

– *A*: electron affinity of a gram atom of non-metal;

– $\Delta_f H$: enthalpy of formation of the solid crystallized from its elements.

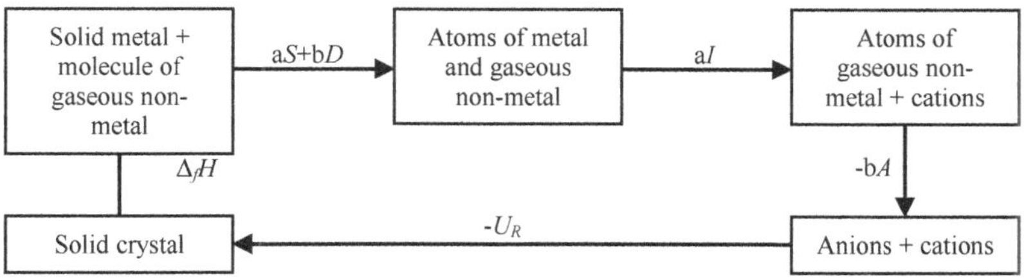

Figure 1.5. *Born–Haber cycle for a compound A_aB_b*

From the cycle shown in Figure 1.5, we deduce the relation:

$$\Delta_f H = \mathrm{a}S + \mathrm{b}D + \mathrm{a}I - \mathrm{b}A - U_{m(R)} \qquad [1.71]$$

If the other values are known, this expression can be used to calculate the crosslink energy $U_{m(R)}$.

Comparison of the direct calculation on the basis of the microscopic model and the result given by the Born–Haber cycle shows that the values overlap at less than 5%. This result is entirely acceptable when we remember the relatively-simple hypotheses of the model, and in particular the assimilation of the ions to non-deformable hard spheres. We can refine the model by introducing deformability of the ions by their polarizability, which leads us to take account of the ion–dipole and dipole–dipole electrostatic interactions.

1.7.3. *Vibrational partition function and internal energy of an ionic solid*

Earlier, when we were looking at molecular solids (see section 1.5.1), we saw the mixture of Debyean and Einsteinian terms in the partition function of the same solid with degrees of vibration of the lattice and degrees of vibration within the molecule. We will now make broader use of this concept, which was developed by Born and Blackman, on the subject of ionic compounds.

To begin with, in the ionic solid, we will construct neutral subsets by associating the minimum number of ions of opposite signs, such as a potassium ion and a chlorine ion to create potassium chloride, a calcium ion and two chlorine ions to make calcium chloride, or a calcium ion and a carbonate ion for calcium carbonate. Such a subset contains s atoms and therefore has $3s$ degrees of freedom – all of them vibrational. The whole of the solid contains N subsets and will therefore have $3Ns$ vibrational degrees of freedom.

Born divides the $3s$ degrees of freedom into two categories, acoustic vibrations and optical vibrations (infrared):

– acoustic vibrations (or phonons), throughout the crystal, have a frequency distribution with a limit frequency. This distribution could be that chosen by Debye (see equation [1.9]) or those proposed by Born or Blackman (Figures 1.4(a) and (b)). These acoustic vibrations give us vibrational terms for the partition function and then for the internal energy, which will be of the same type as those proposed by Debye, in accordance with equations [1.24] for the vibrational partition function and [1.29] for the internal energy;

– optical vibrations, each with its own frequency, which, for the partition function, has a sum of Einsteinian terms such as that given by equation [1.3] and for the internal energy a sum of terms such as those in equation [1.28].

However, in order to express these vibrational partition functions and the corresponding internal energies, for a given solid, we need to know the number of vibrations in each group: how many acoustic vibrations will there be, and then how many optical vibrations?

This distribution depends on a number of factors.

Let us first envisage ionic compounds with simple ions – i.e. one anionic atom and one cationic atom, as is the case in potassium chloride.

If the bond is markedly ionic in nature (say, over 50% ionic), we choose six acoustic vibrations. We then need to ask ourselves whether the two groups of three acoustic vibrations have the same value of the

Debye temperature. If the atoms have similar dimensions, as is the case with potassium chloride, we choose a single Debye temperature; if, however, the ions have very different dimensions – as is the case, for instance, with cesium fluoride or lithium iodide – we choose two values of the Debye temperature. The partition function and the internal energy then contain either one or two Debyean terms, depending on the case.

If the bond is mainly covalent, we choose three acoustic vibrations with a Debye temperature and three optical vibrations with three Einsteinian temperatures. The partition function and the internal energy therefore include a Debyean term and the sum of three Einsteinian terms.

If we now look at more complex ionic compounds, such as calcium fluoride or calcium carbonate, we group together species that are linked by a covalent bond, such as the two fluorine atoms in calcium fluoride, or of the carbonate ion in calcium carbonate. With regard to this latter bond between fluorine and calcium or between carbonate and calcium, we refer to the previous case depending on the covalent nature of the bond. Hence, for calcium fluoride which has nine degrees of freedom, six will be acoustic, with the same Debye temperature for the di-fluorine molecules and calcium, and three will be optical vibrations with three corresponding Einsteinian terms. The partition function (and the internal energy which arises from it) contains a Debyean term and three Einsteinian terms.

For calcium carbonate, which has 15 degrees of freedom, six will be acoustic vibrations with two Debye temperatures and nine degrees of freedom will be optical vibrations, with an Einstein temperature for each one. The partition function (and the resulting internal energy) contains two Debyean terms and nine Einsteinian terms.

In fact, that distribution between acoustic vibrations and optical vibrations is only significant at temperatures close to the Debye temperature. If the temperature is very low (much lower than the lowest Debye temperature), only the corresponding Debye term is predominant; the other terms are negligible. This explains why all the internal energies (and, as we will see in section 1.8.1) all the specific heat capacities at constant volume, tend toward zero, with a T^3 law for the latter.

However, at high temperatures, if the temperature is much higher than the highest of the Debye temperatures, only Einsteinian terms are significant: those for which the Einstein temperature is not too low.

1.8. Description of a metallic solid

Over the course of history, pure metal has been described by a variety of models. The initial model, attributable to Drude, considered the metal to comprise a gas of electrons enveloping positive ions in a constant potential. Drude applies Maxwell–Boltzmann statistics to that electron gas. In fact, as the electrons are fermions, it is most appropriate to apply Fermi–Dirac statistics to them, as Sommerfeld did in his model, still using a constant potential. Unlike with molecules, though, because of their low mass, the electrons cannot be used for the approximation of the classic limit statistics given by:

$$n_i = g_i \exp(-\alpha - \beta\varepsilon_i) \qquad [1.72]$$

However, we will see that it is helpful to apply to them the opposite hypothesis to that condition, i.e.:

$$\exp(-\alpha) >> 1 \qquad [1.73]$$

Sommerfeld's model is perfectly well suited for chemical applications – particularly for determining specific heat capacities – but it is totally unsuited for explaining the electrical properties and the experimental fact that not all energy levels are acceptable in the metal. It was the band theory, developed by Brillouin, which was first able to explain these properties and also the nature of the bond between the metal atoms with the introduction of a periodic field.

We will begin by discussing Sommerfeld's constant-field theory, and then go on to show the modifications made by the periodic field and the grouping of electrons in the metal.

1.8.1. *Sommerfeld's electron perfect gas model*

Thus, we consider that a metal whose volume is V contains a certain N number of electrons that are free to move around. The ratio of the number of free electrons to the number of metal atoms contained in the volume V, for the time being, will be taken as being close to 1 (or perhaps 2 or 3). Only with the band model we will be able to actually calculate that ratio, later on. That gas cloud is comparable to perfect gas made up of electrons placed in a constant average potential, so that they are contained within the volume V. We will apply Fermi–Dirac statistics to that perfect gas, accepting condition [1.73] for now, and verifying it at the end of section 1.8.1.1.

1.8.1.1. *Determination of the coefficient α*

To begin with, we will calculate the coefficient α, which is the Lagrange multiplier relative to the numbers of electrons N. In the knowledge that the other multiplier β always has the same value ($1/\mathrm{k_B}T$) irrespective of the molecular statistics, we can calculate α by using the first method described in section A.1.3 of Appendix 1.

Because the electrons obey Fermi–Dirac statistics, the total number thereof obeys the following law:

$$N = \sum_i n_i \qquad [1.74]$$

and their distribution between the different energy levels obeys:

$$\left(n_i\right)_{FD} = \frac{g_i \exp\left(-\alpha - \beta\varepsilon_i\right)}{1 + \exp\left(-\alpha - \beta\varepsilon_i\right)} \qquad [1.75]$$

Hence, the number of free electrons is given by:

$$N = \sum_i \frac{g_i}{1 + \exp\left(\alpha + \dfrac{\varepsilon_i}{\mathrm{k_B}T}\right)} \qquad [1.76]$$

We have chosen $\varepsilon_0 = 0$ as the origin of the energies for the electrons at rest within the metal.

Consider that the electrons, whose mass is m_e, are contained in a cubic box with side length a and volume $V = a^3$. The kinetic energy, which depends on three quantum numbers (l, m and n) is given by:

$$\varepsilon_i = \frac{\mathrm{h}^2}{8m_e a^2}\left(l^2 + m^2 + n^2\right) \qquad [1.77]$$

For each energy level, there are two corresponding electrons with opposite spin, so the degeneration coefficient is $g_i = 2$. If we consider all the states with the same energy ε_i, the quantum numbers l, m and n have values arranged over the surface of the positive quarter of the sphere defined by:

$$\frac{8m_e\varepsilon_i}{\mathrm{h}^2} = \frac{1}{a^2}\left(l^2 + m^2 + n^2\right) \qquad [1.78]$$

Thus, all the states whose energy is less than or equal to ε_i have the values of their quantum numbers located within the positive quarter of the sphere. The number of points with integer coordinates l, m and n within that positive quarter-sphere is equal to the volume of that quarter-sphere, so:

$$\frac{\pi}{6}\left(\frac{8m_e\varepsilon_i}{\mathrm{h}^2}\right)^{3/2} a^3 = \frac{\pi}{6}\left(\frac{8m_e\varepsilon_i}{\mathrm{h}^2}\right)^{3/2} V \qquad [1.79]$$

V is the volume of the space containing the electrons.

NOTE.– We have chosen to use a cubic sample volume, but it is possible to demonstrate that relation [1.79] applies no matter what the shape of that space.

After the degeneration of two of the electrons, the number of electron states with energy less than ε_i is, clearly, double the number of points within the quarter-sphere, so:

$$n_{\varepsilon \le \varepsilon_i} = \frac{\pi}{3}\left(\frac{8m_e\varepsilon_i}{\mathrm{h}^2}\right)^{3/2} V \quad [1.80]$$

By deriving the above relation, we obtain the number of electron states for which the energy level is between ε and $\varepsilon + \mathrm{d}\varepsilon$:

$$g(\varepsilon) = \frac{\pi}{2}\left(\frac{8m_e}{\mathrm{h}^2}\right)^{3/2} V\varepsilon^{1/2}\,\mathrm{d}\varepsilon \quad [1.81]$$

We will consider all these states to be the degeneration of a single energy state ε, so that:

$$g_i = g(\varepsilon) = \frac{\pi}{2}\left(\frac{8m_e}{\mathrm{h}^2}\right)^{3/2} V\varepsilon^{1/2}\,\mathrm{d}\varepsilon \quad [1.82]$$

By substituting this value of g_i back into relation [1.76] and replacing the sum with an integral, as the states are very close to one another, we obtain the following for the number of free electrons:

$$N = \frac{\pi}{2}\left(\frac{8m_e}{\mathrm{h}^2}\right)^{3/2} V\int_0^\infty \frac{\varepsilon^{1/2}\,\mathrm{d}\varepsilon}{1+\exp\left(\dfrac{\varepsilon}{\mathrm{k_B}T}+\alpha\right)} \quad [1.83]$$

For simplicity's sake, we set the new integration variable x such that:

$$\frac{\varepsilon}{\mathrm{k_B}T} = x \quad [1.84]$$

Relation [1.83] is then written as:

$$N = \frac{\pi}{2}\left(\frac{8m_e\mathrm{k_B}T}{\mathrm{h}^2}\right)^{3/2} V\int_0^\infty \frac{x^{1/2}\,\mathrm{d}x}{1+\exp(x+\alpha)} \quad [1.85]$$

To simplify our expressions, we will introduce an energy level that is characteristic of the solid, known as the *Fermi energy* (or *Fermi-level energy*) as being the maximum value reached by the energy of an electron if the N electrons, two by two, fill all the lower energy levels; thus, according to relation [1.80], this energy must satisfy the condition:

$$\frac{N}{V}=\frac{\pi}{3}\left(\frac{8m_e\varepsilon_F}{\mathrm{h}^2}\right)^{3/2} \qquad [1.86]$$

Therefore, by definition, the Fermi energy is:

$$\varepsilon_F=\frac{\mathrm{h}^2}{8m_e}\left(\frac{3N}{\pi V}\right)^{2/3} \qquad [1.87]$$

By taking account of relation [1.85], we find that the value of this Fermi level is:

$$\left(\frac{\varepsilon_F}{\mathrm{k_B}T}\right)^{3/2}=\frac{3}{2}\int_0^\infty\frac{x^{1/2}\,\mathrm{d}x}{1+\exp(x+\alpha)} \qquad [1.88]$$

Therefore, it is helpful to evaluate the integral I defined by:

$$I=\int_0^\infty\frac{x^{1/2}\,\mathrm{d}x}{1+\exp(x+\alpha)} \qquad [1.89]$$

There is no exact analytical solution for this integral, but we can show that a good approximation is given by a limited expansion:

$$I=-\left(\frac{2\alpha^{1/2}}{3}+\frac{\pi^2\alpha^{-1/2}}{12}\right) \qquad [1.90]$$

Thus, if we substitute the value back into relation [1.88], the Fermi energy becomes:

$$\left(\frac{\varepsilon_F}{\mathrm{k_B}T}\right)^{3/2}\approx-\left(\alpha^{3/2}+\frac{\pi^2\alpha^{1/2}}{8}\right) \qquad [1.91]$$

We can see that, if we focus on the first term in this expansion, an approximation of the Fermi level is given by:

$$\varepsilon_F \approx -\alpha \mathrm{k_B} T \qquad [1.92]$$

Based on this last approximation, which we can put back into equation [1.91] in place of the first term on the right-hand side, we obtain an approximate value of α:

$$\alpha = -\frac{\varepsilon_F}{\mathrm{k_B} T}\left(1 - \frac{\pi^2}{12}\left(\frac{\mathrm{k_B} T}{\varepsilon_F}\right)^2\right) \qquad [1.93]$$

We are going to evaluate an order of magnitude for the coefficient α. In order to do so, we choose one electron per atom of the metal, with these atoms being a few angstroms apart from each other, so the ratio N/V is approximately:

$$\frac{N}{V} \approx 10^{29} m^{-3} \qquad [1.94]$$

Thus, using relation [1.88], we find the following for the Fermi level:

$$\frac{\varepsilon_F}{\mathrm{k_B} T} \approx 1.25 \times 10^{-18}\, joules \qquad [1.95]$$

Thus, at a temperature of around 1000 K, using relation [1.83], we obtain the following value for the coefficient α:

$$\alpha \approx -90 \qquad [1.96]$$

Hence, we can verify that the condition [1.73], which we have simply accepted for simplicity's sake up until now, is indeed respected.

Moreover, we can see that the second term in expression [1.93] has a value of approximately $1/90^2 \approx 10^{-4}$, which is much less than 1, which means that approximation [1.92] is also very appropriate.

1.8.1.2. *Kinetic energy of electrons in the metal*

In order to calculate the kinetic energy of the mobile electrons, we use the relation:

$$E = \sum_i n_i \varepsilon_i \qquad [1.97]$$

To do so, we merely need to introduce the term ε into the integral given in relation [1.83]. Thus, we obtain:

$$E_{kin} = \frac{\pi}{2}\left(\frac{8m_e}{\mathrm{h}^2}\right)^{3/2} V\int_0^\infty \frac{\varepsilon^{3/2}\,\mathrm{d}\varepsilon}{1+\exp\left(\dfrac{\varepsilon}{\mathrm{k_B}T}+\alpha\right)} \qquad [1.98]$$

Using the same change of variables as above (see equation [1.84]), and if we introduce the definition [1.87] of the Fermi energy, we find that:

$$E_{kin} = \frac{3}{2}N\mathrm{k_B}T\left(\frac{\mathrm{k_B}T}{\varepsilon_F}\right)^{3/2} V\int_0^\infty \frac{x^{3/2}\,\mathrm{d}x}{1+\exp(x+\alpha)} \qquad [1.99]$$

Just like the integral I in relation [1.89], the integral in terms of x that appears in equation [1.99] has no analytical solution. We can show, however, that its approximate value can be found by the expansion [1.100]:

$$I' = \int_0^\infty \frac{x^{3/2}\,\mathrm{d}x}{1+\exp(x+\alpha)} \approx -\frac{2}{5}\alpha^{3/2} + \frac{\pi^2}{4}\alpha^{1/2} \qquad [1.100]$$

Taking account of the value of α given by equation [1.96], we find:

$$E_{kin} \approx \frac{3}{5}N\varepsilon_F\left[1+\frac{5\pi^2}{12}\left(\frac{\mathrm{k_B}T}{\varepsilon_F}\right)^2\right] \qquad [1.101]$$

Similarly as for relation [1.91], we can content ourselves with the following relation, with a fairly high degree of accuracy:

$$E_{kin} \approx \frac{3}{5} \mathrm{N_a} \varepsilon_F \qquad [1.102]$$

Thus, we obtain an absolute error of around 0.05 J at 1000 K.

Hence, we can see that for the free electrons in a metal, the kinetic energy is non-null at the temperature of 0 K; this is the consequence of the application of the Fermi–Dirac statistics. Thus, we can no longer define a temperature scale on the basis of the kinetic energy of the free electrons in a metal.

1.8.1.3. *Electrochemical potential of the electrons in the metal and the Fermi energy*

Regardless of the type of statistics used, we can show that the two coefficients α and β, relative to the stresses, respectively, linked to the number of moles and to the energy, are connected to one another by the expression:

$$\mu_{\mathrm{A}} = -\frac{\alpha_{\mathrm{A}}}{\beta} \qquad [1.103]$$

Applied to the electrochemical potential of the electron, this gives us:

$$\tilde{\mu}_{el} = -\mathrm{k_B} T \alpha \qquad [1.104]$$

Thus, by using relation [1.93], the electrochemical potential, still using the electrons at rest as the origin of the energies, is given by:

$$\tilde{\mu}_{el} = \varepsilon_F \left[1 - \frac{\pi^2}{12} \left(\frac{\mathrm{k_B} T}{\varepsilon_F} \right)^2 \right] \qquad [1.105]$$

The application of expression [1.92] gives us the approximate value:

$$\tilde{\mu}_{el} = \varepsilon_F \qquad [1.106]$$

Thus, the electrochemical potential of the free electrons, at the temperature of 0 K, is equal to the Fermi energy of the metal. As approximation [1.92] is very accurate, we can say that this electrochemical potential practically does not vary with temperature.

On the basis of the Fermi energy, we define a temperature known as the *Fermi temperature*, as:

$$T_F = \frac{\varepsilon_F}{\mathrm{k_B}} \qquad [1.107]$$

Let us evaluate an order of magnitude for that Fermi temperature. In view of relation [1.95], we have:

$$\varepsilon_F \cong 1.25 \times 10^{-18}\, joules \qquad [1.108]$$

Thus, by applying definition [1.107], we find the value:

$$T_F \cong 9.10^4\, K \qquad [1.109]$$

Hence, in general, the temperature of the metal for which it remains solid is much lower than the Fermi temperature. Therefore, we can consider that the metal always behaves as though the temperature was 0 K, and we can content ourselves with approximation [1.106]. The Fermi temperature defines the temperature beyond which the effects of Fermi–Dirac statistics begin to manifest themselves. We can write that:

$$\text{If } T << T_F, \text{ then } \mathrm{k_B}T << \tilde{\mu}_{el} \qquad [1.110]$$

NOTE.– We sometimes come across the term *Fermi impulsion*, which is the maximum value of the impulsion of the free electrons at the temperature of absolute zero. To find the expression of it, we only need to write:

$$p_F = \sqrt{2m_e\varepsilon_F} = \left(\frac{3\mathrm{h}^3 N}{8\pi V}\right)^{1/3} \qquad [1.111]$$

1.8.1.4. *Energy distribution of the free electrons*

Based on the distribution function, which is written as follows:

$$\langle dn\rangle = \frac{N}{V}\left(\frac{2\pi m \mathrm{k_B} T}{\mathrm{h}^2}\right)^{3/2} \exp\left(-\frac{p_{x_1}^2 + p_{x_2}^2 + p_{x_3}^2}{2m\mathrm{k_B}T}\right) \mathrm{d}\,p_{x_1}.\mathrm{d}\,p_{x_2}.\mathrm{d}\,p_{x_3} .\mathrm{d}\,x_1.\mathrm{d}\,x_2.\mathrm{d}\,x_3 \quad [1.112]$$

we calculate the average number of free electrons which have the energy ε. It is given at 0 K, in light of expressions [1.92] and [1.106], by one of the following relations:

$$\langle n_\varepsilon\rangle = \frac{1}{1+\exp\left(\dfrac{\varepsilon}{\mathrm{k_B}T}+\alpha\right)} = \frac{1}{1+\exp\left(\dfrac{\varepsilon-\varepsilon_F}{\mathrm{k_B}T}\right)} = \frac{1}{1+\exp\left(\dfrac{\varepsilon-\tilde{\mu}_{el}}{\mathrm{k_B}T}\right)} \quad [1.113]$$

Figure 1.6 shows this distribution at the temperature 0 K. Beyond the Fermi level, there are no more electrons.

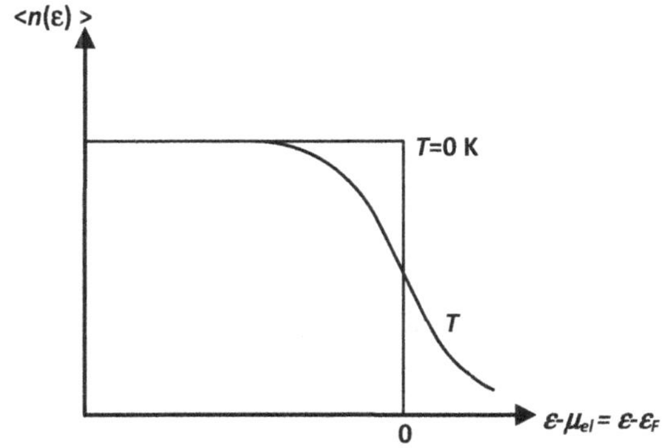

Figure 1.6. *Energy distribution of the free electrons in a metal at 0 K*

At a temperature T, the curve differs only very little from the curve at 0 K, because of the excellent approximation offered by relation [1.92] in respect to the expansion [1.91] and by relation [1.106] in respect to the expansion [1.105], because the temperature of the solid metal is generally far lower than the Fermi temperature.

1.8.1.5. *Contribution of the free electrons to the internal energy of a metal*

On the basis of the electrochemical potential, we can express the function of the electrochemical Helmholtz energy by integrating at constant volume and temperature:

$$\tilde{F}_{el} = \int_0^N \tilde{\mu}_{el} \, \mathrm{d}N \qquad [1.114]$$

By introducing, into that integral, the expression of the electrochemical potential given by relation [1.105], we find:

$$\tilde{F}_{el} = \int_0^N \varepsilon_F \left[1 - \frac{\pi^2}{12} \left(\frac{\mathrm{k_B} t}{\varepsilon_f} \right)^2 \right] \mathrm{d}N \qquad [1.115]$$

We know, from relation [1.87], that the Fermi energy is proportional to $(N/V)^{2/3}$. Thus, for the electrochemical Helmholtz energy, we calculate:

$$\tilde{F}_{el} = N \left[\frac{3}{5} \varepsilon_F - \frac{\pi^2}{4\varepsilon_F} (\mathrm{k_B} T)^2 \right] \qquad [1.116]$$

This expression gives us the electrochemical Helmholtz energy, having chosen $\varepsilon_0 = 0$ as the origin of the energies, for the electrons at rest within the metal, in section 1.8.1.1. In order to compare the energies of the electrons with the contributions of the other terms for the solid, it is helpful to change the origin of that energy by taking as the origin the electrons at rest very far from the metal and very far removed from one another. This introduces a term E_p at a given temperature and volume. This energy represents an average potential energy of the free electrons obtained by averaging the attractions of the positive ions and the other electrons. We suppose that this energy, which is electrostatic in origin, is independent of the temperature but does depend on the volume of the solid, which alters the distances between the ions. If we change the temperature, this energy will only vary by way of the variation in volume – i.e. the thermal expansion.

For a mole of metal, if we let n_{el} denote the number of free electrons per atom of metal, we can rewrite relation [1.116] in the form:

$$\tilde{F}_{el} = n_{el}\mathrm{N_a}\left[-E_p(V) + \frac{3}{5}\varepsilon_F(V) - \frac{\pi^2}{4\varepsilon_F(V)}(\mathrm{k_B}T)^2\right] \qquad [1.117]$$

Based on this expression [1.117] of the Helmholtz energy, which is a characteristic function in our choice of variables, we have access to all the thermodynamic properties, and in particular, we can calculate the contribution of the free electrons to the (electrochemical) internal energy of the metal by using the relation:

$$\tilde{U}_{el} = -T^2 \frac{\partial\left(F_{el}/T\right)}{\partial T} \qquad [1.118]$$

This gives us:

$$\tilde{U}_{el} = \frac{n_{el}\mathrm{N_a}}{T}\left[-E_p(V) + \frac{3}{5}\varepsilon_F(V) + \frac{\pi^2}{4\varepsilon_F(V)}(\mathrm{k_B}T)^2\right] \qquad [1.119]$$

This expression will be used later on when we are calculating the contribution of the free electrons to the molar specific heat capacity of the metal at constant volume (see section 1.8.2.1). However, in order to do that, we need to know the term n_{el}, which is the number of free electrons per atom of the metal. Sommerfeld's model does not provide us with this number, but Brillouin's band theory, or zone theory, can be used to evaluate it.

1.8.2. *The metallic bond and band theory*

As we have just seen, the average potential theory cannot be used to determine the number of free electrons per atom of metal. In order to find it, we need to return to the average potential hypothesis and take account of a periodic potential. To do so, we will recap a few details about metallic bonds.

1.8.2.1. *Origin of energy bands*

In the metal, each of the electrons is subject to the influence of all the nuclei and all the other electrons. In view of the periodic arrangement of the atoms, that potential is periodic, becoming infinite at each nucleus and minimal at the points furthest from the nuclei. We cannot hope to solve the Schrödinger equation for so complex a system. Certain calculations have been performed in specific cases – in particular, by Bloch, Brillouin, Wigner, Seitz and Slater, among others.

In order to gain an understanding of the formation of the bands, we will greatly simplify the system, by considering a one-dimensional solid formed of an infinitely-long line of ions, along which, atomic nuclei are arranged at an equal distance from one another (Figure 1.7). To simplify the problem of quantum mechanics, we suppose that the electrons are classified into two categories;

– the electrons in the outer layer, which are usually the bonding electrons, and are shared in overall orbitals created by the overlapping of the individual orbitals;

– the electrons in the inner layers, which we can assume are not highly affected by the neighboring ions or by the electrons in the bond. Hence, these electrons remain in the vicinity of their respective nuclei.

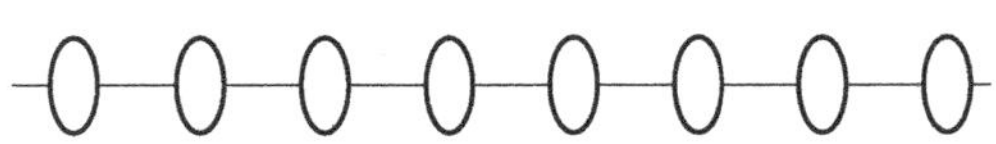

Figure 1.7. *Diagrammatic representation of a one-dimensional metal*

We suppose that the wave function of the bond electrons is independent of the wave function of the ions comprising nuclei and inner-layer electrons.

Thus, it is useful to solve the Schrödinger equation for the valence electrons, which is written as:

$$\frac{\mathrm{h}^2}{8\pi^2 m_e}\Delta\Phi + (E_p - \varepsilon)\Phi = 0 \qquad [1.120]$$

We construct the wave function of the valence electrons by linear combination of the atomic orbitals by adding the atoms in the line one after another, reasoning on the example of the *s* orbitals.

The first atom has an *s* orbital with a certain level of energy. When a second atom is added, its *s* orbital overlaps with that of the previous atom, and forms two molecular orbitals – one bonding and the other antibonding. If we add a third atom, its *s* orbital overlaps with the previous two molecular orbitals to form three new molecular orbitals, and so on. Hence, by adding *n* atoms, we form *s* molecular orbitals, which, as new atoms are added, extends the available energy domain covered by molecular orbitals.

We will solve the Schrödinger equation [1.120] by using the method of addition of the states in the Hückel approximation (see Appendix 2). The secular determinant is written as:

$$\begin{vmatrix} H_{i,i}-\varepsilon & H_{i,j} & 0 & 0 & \ldots & 0 \\ H_i & H_{i,i}-\varepsilon & H_i & 0 & \ldots & 0 \\ 0 & H_i & H_{i,i}-\varepsilon & H_i & \ldots & 0 \\ 0 & 0 & H_i & H_{i,i}-\varepsilon & \ldots & 0 \\ \vdots & \vdots & \vdots & \vdots & \ldots & \vdots \\ 0 & 0 & 0 & 0 & \ldots & H_{i,i}-\varepsilon \end{vmatrix} = 0 \qquad [1.121]$$

$H_{i,i}$ and H_i, respectively, denote the Coulomb integral and the resonance integral between two adjacent atoms.

The determinant theory applied to this triangular determinant yields the solution:

$$\varepsilon_k = H_{i,i} + 2H_i \cos\frac{k\pi}{n+1}, \text{ where } k = 1, 2, \ldots, n \qquad [1.122]$$

When the number n of nuclei is very large, the difference between two energy levels, corresponding to two successive values k and $k + 1$, is very slight, so that we can consider that the levels are continuous, but relation [1.122] gives a solution that is acceptable only if the cosine is between −1 and +1:

$$-1 \leq \cos\frac{k\pi}{n+1} \leq 1 \qquad [1.123]$$

In view of relation [1.122], this gives us the twofold inequality at the conditions on the energy:

$$H_{i,i} - 2H_i \leq \varepsilon_k \leq H_{i,i} + 2H_i \qquad [1.124]$$

This means that only one energy band is allowed, and that the breadth of that band, therefore, is:

$$\Delta\varepsilon_k = 4H_i \qquad [1.125]$$

The energy band thus permitted will be known as an s band (Figure 1.8).

Similarly, if the atoms used have available p orbitals, they will form a p band (Figure 1.8). If the energies in the p band are greater than those of the s band, the p band will be situated above the s band, and between the two, there may be an energy band that is "prohibited" to electrons.

Depending on the case, the bands formed may be spaced a long way apart, close together or even overlapping. Hence, the bands 2s and 2p overlap partially (we will see later on that this overlap accounts for the conductive behavior of alkaline earth metals). Similarly, the bands 3d and 4s overlap, and this explains the conductive properties of the transition metals.

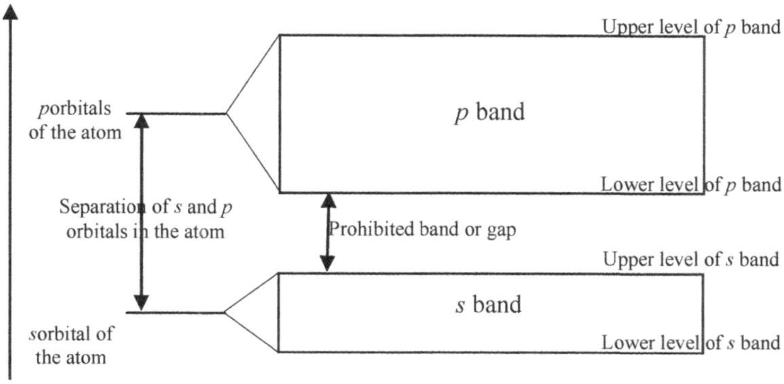

Figure 1.8. *Combination of levels into bands*

1.8.2.2. *Conductors, insulators and semiconductors*

If we look at the Fermi level in the conduction band, we encounter two possible cases:

– in the first case, the Fermi level is included in the conduction band (Figure 1.9, left-hand diagram). A temperature rise enables electrons to pass to a higher level in that band which is still available. Those electrons facilitate conduction, and the solid is cataloged as a conductor. Such is the case, in particular, with metals. The number of excited electrons will be greater when the temperature is higher, but with this rise in temperature, the vibrations of the atoms become more intense, which decreases the electrons' mobility. On balance, the second effect wins out over the first, and the conductivity of metals decreases at higher temperatures;

– in the second case, the Fermi level is identical to the upper level of the conduction band (Figure 1.9, right-hand diagrams). Thus, there are no more levels available in the conduction band to accommodate electrons. The solid is then said to be non-conductive of electricity. In order for a substance to be able to conduct electricity, the electrons must be capable of crossing over a prohibited band and reaching a level in a new authorized band. Thus, two scenarios may arise:

- either the energy jump required to cross the gap is too great in comparison to k_BT, and so the gap cannot be crossed. In this case, we say that the non-conductor is an insulator,

- or the gap is sufficiently narrow, and electrons can cross it, thus freeing up places in the conduction band and facilitating the passage of electrical current. We then say that the non-conductor is a semiconductor. When the temperature increases, the effect due to the increased number of electrons jumping to the highest level prevails over the effect due to the most intense vibrations of the atoms, and the conductivity of semiconductors increases with temperature.

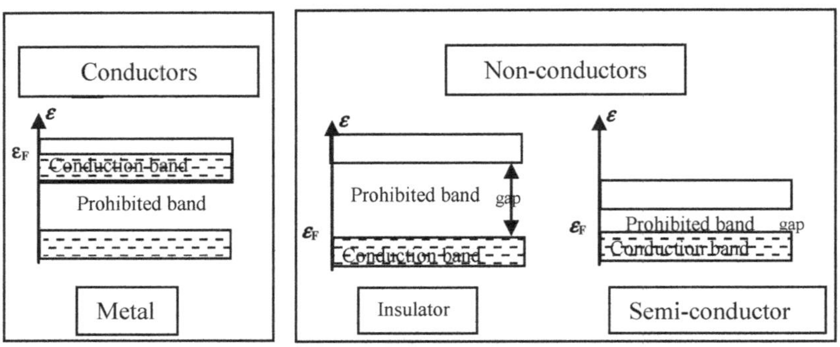

Figure 1.9. *Diagram of bands for metals, insulators and semi-conductors*

1.8.2.3. *Determination of the number N of free electrons*

We will say that the free electrons are the electrons contained in the band with highest energy, which may be completely or partially filled (see Figure 1.10). Hence, the number of free electrons is N and n_e is the ratio of that number N to the number of metal atoms n contained in the volume in question.

The band contains the electrons that are furthest away from the nuclei, and thus we can consider that the amplitude of the periodic potential created by those electrons is slight. Therefore, we can make the approximation that, in the conduction band, the potential is essentially constant. Thus, all the calculations performed in section 1.8.1 are valid within that band, if we take the bottom of the band as the origin of the energies. Thus, the Fermi energy is given by relation [1.87], which is the height, in that band, occupied by electrons at absolute zero.

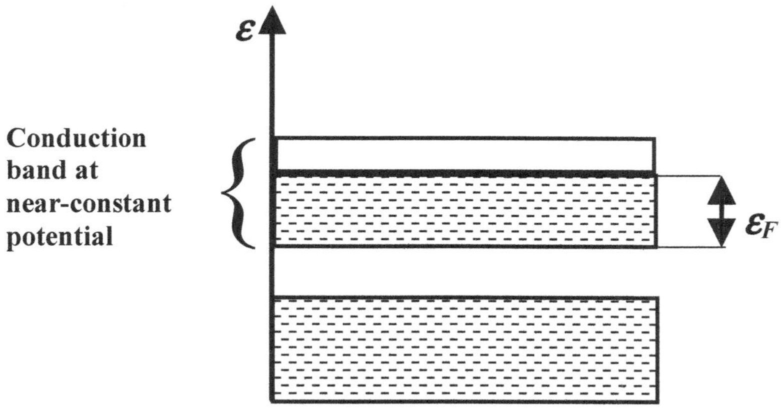

Figure 1.10. *Energy band structure in a metal*

Using this reasoning, it is easy to calculate N. We will now take a look at a few examples.

Let us choose an alkali metal (such as Li, Na, etc.). These atoms have a final electron layer which has one free electron on an s orbital. If the metal contains n atoms, their overlapping will give us $n \times s$ orbitals, and therefore n energy levels in the s band, which can thus accept $2n$ electrons. As there are fewer than $2n$ of them, the conduction band is incomplete, and the Fermi level is within that band. Thus, the ratio n_e is equal to 1, so there is one free electron per metal atom.

Let us now consider an alkaline earth metal (Mg, Ca, etc.). These atoms have a final electron layer containing two electrons on an s orbital. If the metal contains n atoms, then their overlap will give $n \times s$ orbitals, so n energy levels in the s band, which can therefore accommodate $2n$ electrons. Hence, the s band is complete and the Fermi level is at the upper level of that band. Alkaline earth metals should not exhibit metallic behavior; however, experience proves that they are, indeed, metals. This arises from the fact that, in these elements, the s band partially overlaps the p band immediately above it, to form one conduction band, so that the Fermi level actually lies somewhere within that conduction band, because the new band can accommodate more than two electrons per atom (with a maximum of eight if the *s* and *p* bands were precisely tangential to one another). Hence, the ratio n_e is equal to 2, meaning that there are two free electrons per atom of metal. These metals, which have two free electrons per atom, should conduct electricity better than alkali metals. However, this is absolutely not the case, because, at a temperature higher than absolute zero, the number of single electrons, which are responsible for conduction, is lesser in alkaline earth metals than in alkali metals.

We now look at the example of solid fluorine (or any other halogen). Its atoms have a last electron layer comprising two electrons on an *s* orbital and 5 electrons on three equivalent *p* orbitals. If the solid contains *n* atoms, then their overlap will give us $n \times s$ orbitals and $3n \times p$ orbitals, and therefore a total of 4*n* energy levels which can accommodate 8*n* electrons, so the band $(s + p)$ is incomplete, and the Fermi level is included in that band. Halogens should behave like metals, but experience tells us that they are not conductive. This arises from the fact that it is not atoms of fluorine which make up the solid, but rather diatomic F_2 molecules. The combination of those *n*/2 molecules of di-fluorine gives us $n \times s$ orbitals, $2n \times p$ orbitals and *n*/2 orbitals linking two fluorines, and all of those levels, of where there are 7*n*/2, can accept up to 7*n* electrons, which means that the Fermi level is at the highest part of the *sp* band thus constructed, and therefore fluorine is an insulator.

1.8.2.4. *Distribution of energy states and of free electrons at absolute zero*

In order to find the distribution of the electrons, it is important to know the distribution function $g(\varepsilon)$ which, in a solid with volume *V*, gives us the number $g(\varepsilon)$ of states whose energy is between ε and ε+dε. Each state is capable of accommodating two electrons. It is complicated to calculate this distribution. Figure 1.11 shows the distribution calculated by Jones and Mott for the band created by the *s* and *p* bands in a centered cubic crystal.

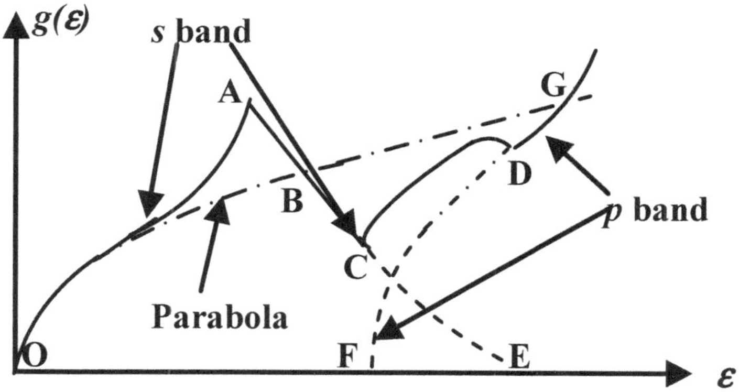

Figure 1.11. *Distribution of states of s and p bands, with overlap, for a centered cubic crystal (data from [FOW 49])*

The curve OABCE represents the distribution in the *s* band, which has the lowest energy (the curve OBG is a parabola). The curve FDG represents the distribution in the start of the *p* band, which is highest. The plot OABCDG shows the distribution resulting from the partial overlap of the *s* and *p* bands.

Figure 1.12 shows the distribution of the energy levels and electron filling in four cases at the temperature of absolute zero. Cases (*a*) and (*b*) correspond respectively to mono- and divalent metals (alkali metals and alkaline earth metals found in section 1.8.2.3). These are conductive metals, in which the *s* and *p* bands partly overlap. In case (*a*), the electrons do not completely fill the *s* band; in case (*b*), they completely fill the *s* band, but the *p* band is available because of the partial overlap. Case (*c*) corresponds to the case of a non-insulating semiconductor, wherein the bands do not overlap. The lower band is full but the upper band is very close to the former. Case (*d*) corresponds to an insulator, where the bands do not overlap. The lower band is full, and the upper band is very far removed and is empty.

Thus, we find (*a*), (*b*) and (*c*) – the three cases of conduction illustrated in Figure 1.9.

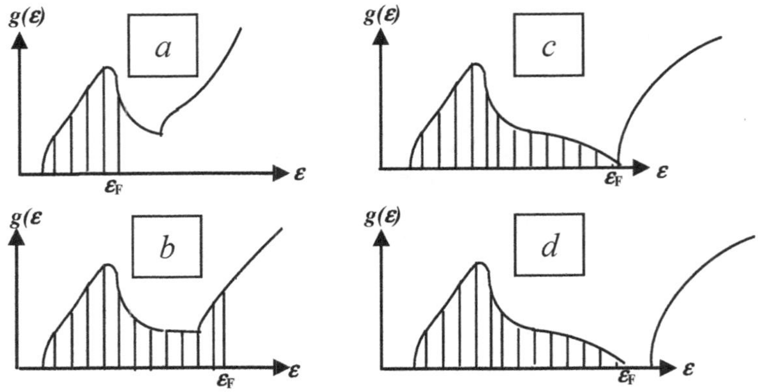

Figure 1.12. *Occupation of states in a) a monovalent metal; b) a divalent metal; c) a semi-conductor; d) an insulator*

1.9. Molar specific heat capacities of crystalline solids

We can determine the molar specific heat capacities at constant volume by derivation of the molar internal energy, using the relation:

$$C_V = \left(\frac{\partial U_m}{\partial T}\right)_V \quad [1.126]$$

We know that depending on the solids, the molar internal energy is the sum of several contributions – e.g. the sum shown in relation [1.35]. By application of relation [1.126], we find that the overall specific heat capacity will also be the sum of various contributions, but only the contributions of the internal energy, which are temperature-dependent, give us a term for the contribution to the overall specific heat capacity.

1.9.1. *Contribution of the vibrational energy to the specific heat capacity at constant volume*

The contribution of the vibrational motions to the internal energy is found in all crystalline solids, and is always temperature-dependent. We have found two types of contributions:

– the so-called Einsteinian contribution, with a unique vibration, mainly applicable to an optical vibration;

– the distribution of $3N$ acoustic vibrations in Debye's model and its derivatives.

We will now calculate the contribution corresponding to each of these models to the specific heat capacity at constant volume.

1.9.1.1. *Case of a unique vibration in Einstein's model*

By combining expressions [1.29] and [1.126], we obtain the contribution of a unique vibration to the specific heat capacity at constant volume:

$$C_{V(v)} = 3\mathrm{R}\left(\frac{\Theta_E}{T}\right)^2 \frac{\exp\left(\frac{\Theta_E}{T}\right)}{\left[\exp\left(\frac{\Theta_E}{T}\right)-1\right]^2} \qquad [1.127]$$

We often adopt the function $\mathrm{E}(T/\Theta_E)$, known as *Einstein's function*, which is defined by:

$$\mathrm{E}\left(\frac{T}{\Theta_E}\right) = \left(\frac{\Theta_E}{T}\right)^2 \frac{\exp\left(\frac{\Theta_E}{T}\right)}{\left[\exp\left(\frac{\Theta_E}{T}\right)-1\right]^2} \qquad [1.128]$$

The corresponding specific heat capacity will thus be written:

$$C_{V(v)} = 3\mathrm{RE}\left(\frac{T}{\Theta_E}\right) \qquad [1.129]$$

At high temperature, the molar specific heat capacity is reduced to:

$$C_{V(v)} \cong 3\mathrm{R} \qquad [1.130]$$

This is the Dulong and Petit law, which we can deduce from the kinetic theory of gases, noting that $3N$ vibrational degrees of freedom correspond to $6N$ quadratic terms.

At low temperature, the molar specific heat capacity tends toward zero.

The Einstein temperature is obtained on the basis of the infrared spectra which enable us to determine the vibration frequencies of the links and then apply the expression $\Theta_E = \frac{h\nu}{k_B}$.

1.9.1.2. *Case of Debye's acoustic vibration distribution*

By combining relations [1.30] and [1.126], we obtain the contribution of an acoustic vibration distribution, according to Debye, to the specific heat capacity at constant volume, so:

$$C_{V(v)} = 12\mathrm{R}\left(\frac{T}{\Theta_D}\right)^3 \int_0^{\Theta_D/T} \frac{x^3}{\exp(x)-1}\mathrm{d}x \qquad [1.131]$$

We often posit the function $\mathrm{D}(T/\Theta_E)$, known as *Debye's function*, defined by:

$$\mathrm{D}(T/\Theta_D) = 4\left(\frac{T}{\Theta_D}\right)^3 \int_0^{\Theta_D/T} \frac{x^3}{\exp(x)-1}\mathrm{d}x \qquad [1.132]$$

The corresponding specific heat capacity is then written:

$$C_{V(v)} = 3\mathrm{R}\,\mathrm{D}\left(\frac{T}{\Theta_D}\right) \qquad [1.133]$$

At high temperature, the molar specific heat capacity is reduced to:

$$C_{V(v)} \cong 3\mathrm{R} \qquad [1.134]$$

We see the Dulong–Petit law.

At low temperature ($T \ll \Theta_D$), relation [1.26] gives the so-called "T^3" law, for the specific heat capacity:

$$C_{V(v)} = \frac{12}{5} R\pi^4 \left(\frac{T}{\Theta_D} \right)^3 \qquad [1.135]$$

Figure 1.13 compares the curve given by the Einsteinian relation [1.129] and that given by Debye's law [1.133]. We can see a difference between the two curves – particularly at low temperature.

The Debye temperature can be measured in a variety of ways. The most common are:

– evaluation on the basis of the whole curve $C_v(T)$;

– evaluation on the basis of the values of the specific heat capacities obtained at low temperature (less than $\Theta_D/12$), and application of the T^3 law [1.135];

– the Debye temperature can also be deduced from elastic data in the vicinity of 200 K.

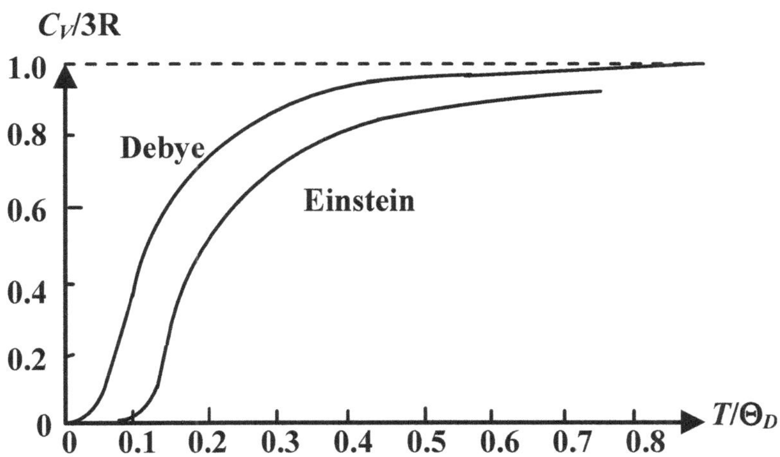

Figure 1.13. *Comparison of the curves of the Einstein and Debye contributions for the specific heat capacity*

Figure 1.14 gives a few of the values that are used. For example, for copper, the values obtained by the different methods all fall within the range 310–340 K.

If we choose a wavelength distribution other than Debye's, then the laws are obviously changed. For example, the Born–von Karmann distribution (Figure 1.4(a)) gives us a law combining the Einsteinian and Debyean terms, as follows:

$$C_{V(v)} = \frac{3}{2}\mathrm{R}\left[\mathrm{E}\left(\frac{\gamma T}{\Theta_m}\right) + \mathrm{D}\left(\frac{0{,}8T}{\Theta_m}\right)\right] \qquad [1.136]$$

where γ and Θ_m being two constants, which we adjust in relation to the experimental data.

Choosing to use Blackman's distribution (Figure 1.4(b)), Nernst and Lindemann put forward the following law:

$$C_{V(v)} = \frac{3}{2}\mathrm{R}\left[\mathrm{E}\left(\frac{T}{\Theta}\right) + \mathrm{E}\left(\frac{2T}{\Theta}\right)\right] \qquad [1.137]$$

We are going to make use of these contributions of the vibrations to the construction of the specific heat capacity at constant volume for different types of solids.

1.9.2. *Specific heat capacity of an atomic solid at constant volume*

An atomic solid has an energy given by a relation similar in form to equation [1.28], which involves a temperature-dependent vibrational term, which gives us a vibrational contribution to the specific heat capacity of the form:

$$C_{V(v)} = 3\mathrm{R}\,\mathrm{D}\left(\frac{T}{\Theta_D}\right) \qquad [1.138]$$

A priori, the vibrational term is the only one in equation [1.28] which is temperature-dependent. However, we need to take a slightly closer look at the electron contribution – particularly in the wake of our study of section 1.7, and distinguish conductive atomic solids from non-conductive solids.

1.9.2.1. *Case of conductors*

Conductive solids have free levels for the electrons above the Fermi level in the conduction band. Thus, we distinguish between:

– electrons in the lower bands, which are not excited, whose internal energy does not depend on the temperature (as envisaged in relation [1.28]) and which make no contribution to the specific heat capacity;

– electrons which can be described as "free", located at the temperature 0 K in the conduction band below the Fermi level. At temperature T, the internal energy of these free electrons is given by relation [1.119], which is a function of the temperature. By applying expression [1.126], we can deduce a contribution of the free electrons to the specific heat capacity, given by:

$$C_{V(el)} = \frac{4\chi}{3^{2/3}} \frac{\mathrm{R}\pi^{8/3} V^{2/3}}{\mathrm{h}^2 \mathrm{N}_\mathrm{a}^{2/3}} T \qquad [1.139]$$

Thus, strictly speaking, the specific heat capacity of a metal at constant volume will be the sum of two contributions: one from the vibration of the ions and the other from the free electrons, so by applying the contributions [1.138] and [1.139]:

$$C_V = \frac{4\chi}{3^{2/3}} \frac{\mathrm{R}\pi^{8/3} V^{2/3}}{\mathrm{h}^2 \mathrm{N}_\mathrm{a}^{2/3}} T + 3\mathrm{R}\,\mathrm{D}\left(\frac{T}{\Theta_D}\right) \qquad [1.140]$$

Based on this general relation, we can formulate two observations:

1) The first deals with the relative orders of magnitude of the two terms in addition [1.140]. The electronic term would be order of 5×10^{-4} TJ/mole.K^{-1}, which means it is negligible at normal temperatures. It only becomes significant at around 1 K. This was confirmed by Keesom and Kok, who measured such a contribution of 2.5×10^{-4} T for copper, which corresponds to a contribution of

0.033 J/mole.K^{-1} at 133 K, as compared to a vibrational contribution of 16.2 J/mole.K^{-1}.

2) The second observation pertains to the influence of the temperature. The same two authors observed that in the vicinity of absolute zero, the value of the specific heat capacity is greater than that given by the extrapolation of the T^3 law given by equation [1.135]. The capacity is thus written as:

$$C_V = \frac{4\chi}{3^{2/3}} \frac{\mathrm{R}\pi^{8/3} V^{2/3}}{\mathrm{h}^2 \mathrm{N}_\mathrm{a}^{2/3}} T + \frac{12}{5} \mathrm{R}\pi^4 \left(\frac{T}{\Theta_D} \right)^3 \quad [1.141]$$

At the temperature of 1 K, the T term could even become predominant.

1.9.2.2. *Case of insulating materials*

In the case of insulating materials, the electrons in the conduction band do not have the opportunity to gain energy by an increase in temperature, as the gap to be crossed is very large indeed, so the electron contribution will be null. However, as we noted in the case of fluorine, the crystal must be considered to be a crystal of diatomic molecules. This being the case, each molecule F_2 has six degrees of freedom. Three will be degrees of acoustic vibrations, arising from the Debye term, and three will be optical vibrations, each arising from an Einsteinian term. Thus, the specific heat capacity contains only a vibrational contribution, formed of four terms in accordance with the following (for a mole of fluorine or ½ a mole of difluorine):

$$C_V = \frac{3\mathrm{R}}{2} \mathrm{D}\left(\frac{T}{\Theta_D} \right) + \frac{\mathrm{R}}{2} \sum_{s=4}^{s=6} \mathrm{E}_\mathrm{s} \left(\frac{T}{\Theta_{E(s)}} \right) \quad [1.142]$$

If the three Einsteinian terms have very similar frequencies, then the relation can be simplified to:

$$C_V = \frac{3\mathrm{R}}{2} \left[\mathrm{D}\left(\frac{T}{\Theta_D} \right) + \mathrm{E}\left(\frac{T}{\Theta_E} \right) \right] \quad [1.143]$$

In fact, in general, the Debye temperature is much lower than the Einstein temperature, so we can divide the temperature scale into three regions:

– if $T<<\Theta_D$, the Debye term is predominant, the Einstein term can be ignored and we find the T^3 law around the temperature of absolute zero;

– if $T>>\Theta_D$, the Einstein term is predominant, and Debye's can be ignored.

Only if the temperature lies between those two characteristic temperatures we do need to apply relation [1.143].

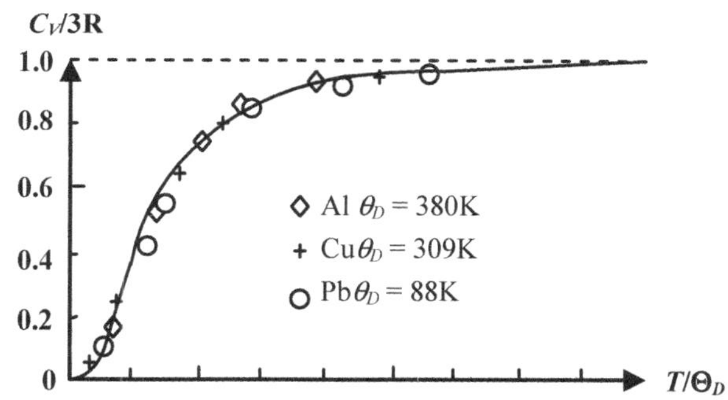

Figure 1.14. *Debye curve and specific heat capacities at constant volume for a number of atomic solids*

NOTE.– In the case of semiconductors, the situation is the same as for insulators, but there is the possibility of populating the band above the Fermi level by increasing the temperature, which would introduce an electron term, as a function of the temperature in the internal energy and therefore an electron contribution to the specific heat capacity. However, the breadth of the gap to be crossed is such that few electrons can make that jump at low temperature, and therefore the corresponding term will always be negligible in comparison to the vibrational term.

Figure 1.14 shows that Debye's curve corresponds closely with the variations in the $C_V/3R$ ratio for a number of metals. Note that the curve is only plotted above a certain temperature, above the temperature where the electronic term becomes significant.

1.9.3. *Specific heat capacity of a molecular or ionic solid at constant volume*

For molecules, as for ionic solids after grouping into neutral entities (see section 1.7.3), the internal energy is given by the sum such as [1.35], wherein only the vibrational term varies with temperature.

However, as we have seen in section 1.7.3, the vibrations are divided into acoustic vibrations by multiples of three and arising from the Debyean terms and optical vibrations, whose number is complementary to the former and which all result from an Einsteinian term. Hence, the specific heat capacity takes the following form:

$$C_V = 3\mathrm{R}\left[\sum_{\mathrm{ac}} \mathrm{D}\left(\frac{T}{\Theta_D}\right) + \sum_{\mathrm{opt}} \mathrm{E}\left(\frac{T}{\Theta_E}\right)\right] \quad [1.144]$$

In general, at low temperatures, only the Debye terms need to be taken into account, whereas at high temperatures, the Einstein terms are largely predominant. This means that the T^3 extrapolation to absolute zero is correctly observed, because only the Debye term remains.

1.9.4. *Conclusion as to the specific heat capacity of a crystalline solid*

In conclusion, it can very often be admitted that, outside of metals, the specific heat capacity varies with temperature like Debye's law at low temperatures, with a law proportional to T^3 at very low temperatures (below 40 K), while Einstein's law performs better at higher temperatures. For metals, caution needs to be exercised when we come very close to absolute zero: a law proportional to T is better than the T^3 law.

At very high temperatures, all laws tend toward the Dulong–Petit law, meaning that the specific heat capacity at constant volume becomes temperature-independent, obeying relation [1.130].

Macroscopic modeling always gives us the specific heat capacity at constant volume. We can switch to the specific heat capacity at constant pressure by using the relation:

$$C_P - C_V = T\left(\frac{\partial P}{\partial T}\right)_V \left(\frac{\partial V}{\partial T}\right)_P \qquad [1.145]$$

Using the thermomechanical coefficients (see section 1.9.1), we construct expression [1.146], which links the difference between the specific heat capacities at constant pressure and volume to the expansion coefficient β, and to the compressibility coefficient χ_T :

$$C_P - C_V = \frac{V\beta^2}{\chi_T} \qquad [1.146]$$

This very general relation for solids is the equivalent of relation [1.145], but is easier to use for gases on the basis of the state equation.

1.10. Thermal expansion of solids

The thermal expansion of a solid is the variation in that solid's dimensions when its temperature is changed. A few exceptions aside, dimensions generally increase when the temperature rises. This expansion is due to the anharmonicity of the molecules' vibrations.

1.10.1. *Expansion coefficients*

On the macroscopic level, the expansion is characterized by the linear and cubic expansion coefficients.

1.10.1.1. *Linear expansion coefficient*

In a direction in space, the increase of a length under the influence of temperature is defined at a given pressure by the linear expansion coefficient at constant pressure α:

$$\alpha = \frac{1}{l}\left(\frac{\partial l}{\partial T}\right)_P \qquad [1.147]$$

We often express that coefficient in relation to the l_0 at the temperature of 0°C, defining a standard linear expansion coefficient α_0 as:

$$\alpha_0 = \frac{1}{l_0}\left(\frac{\partial l}{\partial T}\right)_P \quad [1.148]$$

Table 1.3 gives the value of the linear expansion coefficient for a number of substances at standard temperature.

This linear expansion coefficient can be measured either by using a dilatometer or by shift in X-ray diffraction lines as a function of the temperature.

Substance	α (10^{-6} C^{-1})	Substance	α (10^{-6} C^{-1})
Aluminum	22.38	Brass	18.5
Copper	16.70	Invar	1
Iron	11.70	Glass	7
Lead	27.26	Pyrex glass	3
Tantalum	6.46	Quartz	0.55
Tungsten	4.28	Porcelain	3
Zinc	35.40		

Table 1.3. *Linear expansion coefficients of a number of substances*

1.10.1.2. *Thermal expansion tensor*

In anisotropic solids, the expansion coefficient depends on the direction, so to describe the expansion we use a second-order symmetrical tensor, which, in the case of a triclinic solid, has six expansion coefficients:

$$\begin{bmatrix} \alpha_{11} & \alpha_{12} & \alpha_{13} \\ \alpha_{21} & \alpha_{22} & \alpha_{23} \\ \alpha_{31} & \alpha_{32} & \alpha_{33} \end{bmatrix}$$

Because this tensor must be symmetrical, we have:

$$\alpha_{12} = \alpha_{21};\ \alpha_{13} = \alpha_{31} \text{ and } \alpha_{23} = \alpha_{32}$$

With an orthorhombic solid, the tensor is diagonal, the three terms α_{12}, α_{12} and α_{12} have a value of zero, and the three diagonal terms α_{12} , α_{12} and α_{12} give the expansion along the three axes *a*, *b* and *c* of the material.

The eigenvalues of that tensor are the three primary expansion coefficients α_1, α_2 and α_3.

1.10.1.3. *Cubic expansion coefficient (or coefficient of relative volume increase)*

To characterize changes in volume under the influence of temperature, we define a volume expansion coefficient for a level of pressure maintained as constant, thus:

$$\beta = \frac{1}{V}\left(\frac{\partial V}{\partial T}\right)_P \qquad [1.149]$$

Sometimes that coefficient is expressed in relation to the volume V_0 occupied at the temperature of 0°C:

$$\beta_0 = \frac{1}{V_0}\left(\frac{\partial V}{\partial T}\right)_P \qquad [1.150]$$

The volume expansion coefficient is obtained by "plotting" the expansion tensor – i.e. by the sum of the three primary coefficients – but also, because the plot of a square matrix is invariant in a changed system of coordinates, by the sum of the three diagonal terms in the thermal expansion tensor, which gives us the relations:

$$\beta = \alpha_1 + \alpha_2 + \alpha_3 = \alpha_{11} + \alpha_{22} + \alpha_{33} \qquad [1.151]$$

NOTE.– In isotropic media, the linear expansion coefficient is the same in all three directions in space, so:

$$\alpha_1 = \alpha_2 = \alpha_3 = \alpha$$

Thus:

$$\beta = 3\alpha \qquad [1.152]$$

1.10.1.4. *Relation between the thermomechanical coefficients*

We define the isothermal compressibility coefficient at constant temperature T by the relation:

$$\chi_T = -\frac{1}{V}\left(\frac{\partial V}{\partial P}\right)_T \qquad [1.153]$$

The minus sign is introduced because the true intensive variable, conjugate to the volume, is the opposite of the pressure: $-P$.

We also define the coefficient of pressure increase at constant volume by the relation:

$$\delta = \frac{1}{P}\left(\frac{\partial P}{\partial T}\right)_V \qquad [1.154]$$

If, for the phase in question, there is an equation of state such as $f(P, V, T) = 0$, we have the relation:

$$\beta = \delta\chi_T P \qquad [1.155]$$

Indeed, for that state function, we can write:

$$\delta\chi_T P = -\frac{1}{P}\frac{\dfrac{\partial f}{\partial P}}{\dfrac{\partial f}{\partial T}} \cdot \frac{1}{V}\frac{\dfrac{\partial f}{\partial V}}{\dfrac{\partial f}{\partial P}} P = -\frac{1}{V}\frac{\dfrac{\partial f}{\partial V}}{\dfrac{\partial f}{\partial T}} = \frac{1}{V}\frac{\partial V}{\partial T} = \beta$$

1.10.2. *Origin of thermal expansion in solids*

We will now show that the thermal expansion can be attributed to the vibrations of the atoms at the nodes of the crystalline lattice.

For this purpose, we will use the following values for the definite integrals:

$$\int_{-\infty}^{+\infty} \exp - x^2 \, \mathrm{d}x = \sqrt{\pi} \qquad [1.156]$$

$$\int_{-\infty}^{+\infty} x \exp - x^2 \, \mathrm{d}x = 0 \qquad [1.157]$$

$$\int_{-\infty}^{+\infty} x^3 \exp - x^2 \, \mathrm{d}x = 0 \qquad [1.158]$$

$$\int_{-\infty}^{+\infty} x^4 \exp - x^2 \, \mathrm{d}x = \frac{3\sqrt{\pi}}{4} \qquad [1.159]$$

To begin with, we will consider a harmonic vibration. In conventional mechanics, the curve that gives the potential energy of the harmonic oscillator as a function of the length of the spring is a parabola (Figure 1.15(a)). On that parabola, an increase in temperature results in a rise of the potential energy, at various temperatures T_1, T_2, T_3 and T_4 shown in the figure. When the temperature is raised, we can clearly see from the figure that the average position of the end of the spring does not change, and therefore this average position follows a vertical line. This means that no expansion is observed with this increase in temperature. This can be demonstrated numerically.

The equation for the curve of the parabola is written as:

$$\varepsilon - \varepsilon_0 = a(r - r_0)^2 = ax^2 \qquad [1.160]$$

Let us calculate the average stretching of the spring $<r - r_0>$ by applying the definition of the average of a value Q:

$$<Q> = \frac{\sum_{Accessible\ states} Q}{\sum_{Accessible\ states} 1} \qquad [1.161]$$

When we apply this formula to the average stretching, by using Boltzmann statistics, we find:

$$\left\langle r - r_0 \right\rangle = \frac{\int_{-\infty}^{+\infty} x \exp - \frac{\varepsilon - \varepsilon_0}{\mathrm{k_B} T} \mathrm{d}x}{\int_{-\infty}^{+\infty} \exp - \frac{\varepsilon - \varepsilon_0}{\mathrm{k_B} T} \mathrm{d}x} = \frac{\int_{-\infty}^{+\infty} x \exp - \frac{a x^2}{\mathrm{k_B} T} \mathrm{d}x}{\int_{-\infty}^{+\infty} \exp - \frac{a x^2}{\mathrm{k_B} T} \mathrm{d}x} = 0 \qquad [1.162]$$

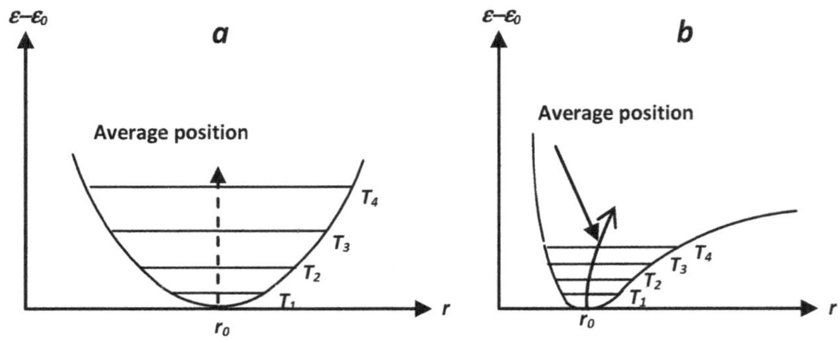

Figure 1.15. *Potential energy curves for a) the harmonic oscillator, and b) an anharmonic oscillator*

We can see from the value of the integral [1.157] that the stretching is zero. Thus, we confirm the approximation of a harmonic oscillator which does not allow for thermal expansion.

Now let us consider an anharmonic oscillator, the potential energy for which is plotted by the curve in Figure 1.15(b). We can clearly see that this shape of curve allows for the breaking of the link. However, we can also see that if the temperature increases, the average position of the end of the spring follows a curve that inclines to the right, thus showing an expansion. We can demonstrate this numerically.

To the equation for the parabola, we add a cubic term to express the anharmonicity of the potential energy curve. Thus, we adopt a position that is fairly close to the minimum – i.e. at low temperatures. Hence, the energy can be written as:

$$\varepsilon - \varepsilon_0 = a(r - r_0)^2 + b(r - r_0)^3 = ax^2 + bx^3 \qquad [1.163]$$

Using Boltzmann statistics like in equation [1.162] but with the new formula for the energy given by relation [1.163], we obtain:

$$\langle r - r_0 \rangle = \frac{\int_{-\infty}^{+\infty} x \exp -\frac{ax^2}{\mathrm{k_B}T} \exp -\frac{bx^3}{\mathrm{k_B}T} \mathrm{d}x}{\int_{-\infty}^{+\infty} \exp -\frac{ax^2}{\mathrm{k_B}T} \exp -\frac{bx^3}{\mathrm{k_B}T} \mathrm{d}x} \qquad [1.164]$$

The exponential term of anharmonicity, which is deemed to be low, is replaced by the development limited to the first two terms:

$$\exp -\frac{bx^3}{\mathrm{k_B}T} \approx 1 - \frac{bx^3}{\mathrm{k_B}T} \qquad [1.165]$$

By substituting this value back into equation [1.164], we find:

$$\langle r - r_0 \rangle = \frac{\int_{-\infty}^{+\infty} x \exp -\frac{ax^2}{\mathrm{k_B}T} \left(1 - \frac{bx^3}{\mathrm{k_B}T}\right) \mathrm{d}x}{\int_{-\infty}^{+\infty} \exp -\frac{ax^2}{\mathrm{k_B}T} \left(1 - \frac{bx^3}{\mathrm{k_B}T}\right) \mathrm{d}x} \qquad [1.166]$$

The numerator and the denominator in this fraction can both be expanded, giving us:

$$\langle r - r_0 \rangle = \frac{\int_{-\infty}^{+\infty} x \exp -\frac{ax^2}{\mathrm{k_B}T} - \int_{-\infty}^{+\infty} \frac{bx^4}{\mathrm{k_B}T} \exp -\frac{ax^2}{\mathrm{k_B}T} \mathrm{d}x}{\int_{-\infty}^{+\infty} \exp -\frac{ax^2}{\mathrm{k_B}T} - \int_{-\infty}^{+\infty} \frac{bx^3}{\mathrm{k_B}T} \exp -\frac{ax^2}{\mathrm{k_B}T} \mathrm{d}x} \qquad [1.167]$$

We recognize the different definite integrals whose values are given in equations [1.156], [1.157], [1.158] and [1.159]. If we substitute in those values, the stretching becomes:

$$< r - r_0 > \; = -\frac{3b}{4a^2} \mathrm{k_B} T \qquad [1.168]$$

The expansion coefficient can thus be obtained on the basis of its definition:

$$\alpha = \frac{1}{r_0} \frac{\mathrm{d}\langle r - r_0 \rangle}{\mathrm{d}T} = -\frac{3b}{4a^2 r_0} \mathrm{k_B} \qquad [1.169]$$

We can see that simply introducing an anharmonic term into the equation for the potential energy curve is sufficient to take account of the existence of linear expansion.

Unfortunately, our model gives us a linear expansion coefficient that is temperature-independent. This result would be kept if a larger number of terms were added to the potential energy equation in relation [1.163]. However, experience tells us that α is dependent on the temperature.

In order to achieve a correct representation of the expansion, we will approach the problem using quantum mechanics.

1.10.3. *Quantum treatment of thermal expansion. Grüneisen parameter*

To begin with, note that by using relations [1.152], [1.154] and [1.155], we can write:

$$\alpha = \frac{\chi_T}{3} \left(\frac{\partial P}{\partial T} \right)_V \qquad [1.170]$$

In order to calculate the linear expansion coefficient, we first need to calculate the thermodynamic coefficient $(\partial P / \partial T)_V$. For this purpose, we will work with the variables V and T.

We will consider our solid to be a collection of quantum oscillators, phonons, each of which has its own natural frequency of vibration ν_i.

The partition function relative to such an oscillator is:

$$z_i = \frac{\exp - \dfrac{\mathrm{h}\nu_i}{2\mathrm{k_B}T}}{1 - \exp - \dfrac{\mathrm{h}\nu_i}{\mathrm{k_B}T}} \qquad [1.171]$$

The characteristic function with variables T, V is the Helmholtz energy. Thus, for each phonon, we can write its contribution to the Helmholtz energy in the form:

$$F_i = -\mathrm{k_B}T \ln z_i = \frac{\mathrm{h}\nu_i}{2} + \mathrm{k_B}T \ln\left[1 - \exp - \frac{\mathrm{h}\nu_i}{\mathrm{k_B}T}\right] \qquad [1.172]$$

Similarly, the contribution of that phonon to the internal energy will be:

$$U_i = -\frac{\partial \ln z_i}{\partial\left(\dfrac{1}{\mathrm{k_B}T}\right)} = \frac{\mathrm{h}\nu_i}{2} + \frac{\mathrm{h}\nu_i}{\exp\dfrac{h\nu_i}{\mathrm{k_B}T} - 1} \qquad [1.173]$$

The anharmonic effect of the oscillations will be taken into account by considering that the vibration frequency of the phonon is a function of the volume.

As the pressure is the derivative of the Helmholtz energy in relation to the volume, the contribution of our phonon to that pressure will be:

$$P_i = -\left(\frac{\partial F_i}{\partial V}\right)_T = -\left[\frac{\mathrm{h}\partial \nu_i}{2\partial V} + \frac{\frac{\partial \nu_i}{\partial V}\mathrm{h}\exp-\frac{\mathrm{h}\nu_i}{\mathrm{k_B}T}}{1-\exp-\frac{\mathrm{h}\nu_i}{\mathrm{k_B}T}}\right] \quad [1.174]$$

From this, we deduce:

$$P_i = -\frac{\partial \nu_i}{\partial V}\left[\frac{\mathrm{h}}{2} + \frac{\mathrm{h}}{\exp\frac{\mathrm{h}\nu_i}{\mathrm{k_B}T}-1}\right] = -\frac{U_i}{\nu_i}\frac{\partial \nu_i}{\partial V} \quad [1.175]$$

Additionally, we can always write:

$$\frac{\partial \nu_\mathrm{i}}{\partial V} = \frac{\nu_\mathrm{i}}{V}.\frac{V}{\nu_\mathrm{i}}.\frac{\partial \nu_\mathrm{i}}{\partial V} = \frac{\nu_\mathrm{i}}{V}.\frac{\partial \ln \nu_\mathrm{i}}{\partial \ln V} \quad [1.176]$$

We define a value γ_i by the expression:

$$\gamma_i = -\frac{\partial \ln \nu_\mathrm{i}}{\partial \ln V} \quad [1.177]$$

This value is called the *Grüneisen factor* of the phonon *i*.

The pressure due to all the phonons, in view of relations [1.176] and [1.177], and by adding relation [1.175] for all the phonons, will be written as:

$$P = \sum_i P_i = \sum_i \gamma_i \frac{U_i}{V} \quad [1.178]$$

We derive this pressure in relation to temperature, at constant volume. In view of expression [1.126], we obtain:

$$\left(\frac{\partial P}{\partial T}\right)_V = \frac{1}{V}\sum_i \gamma_i \left(\frac{\partial U_i}{\partial T}\right)_T = \frac{1}{V}\sum_i \gamma_i C_{Vi} \quad [1.179]$$

By substituting that derivative back into equation [1.170], the expansion coefficient is written as:

$$\alpha = \frac{\chi_T}{3V}\sum_i \gamma_i C_{Vi} = \frac{\chi_T}{3V}\frac{\sum_i \gamma_i C_{Vi}}{\sum_i C_{Vi}}\sum_i C_{Vi} \qquad [1.180]$$

We define the value $\gamma(T,V)$, called the *Grüneisen parameter*, by the expression:

$$\gamma(T,V) = \frac{\sum_i \gamma_i C_{Vi}}{\sum_i C_{Vi}} \qquad [1.181]$$

NOTE.– It is important not to confuse the Grüneisen factor γ_i with the Grüneisen parameter $\gamma(T,V)$.

Thus, the expansion coefficient will be:

$$\alpha = \frac{\chi_T}{3}\gamma(T,V)C_V \qquad [1.182]$$

As the isothermal compressibility χ_T and the specific heat capacity at constant volume C_V are both values which are always positive, the Grüneisen parameter has the same sign as the expansion coefficient – i.e. positive in the majority of cases and negative in the few cases of contraction with increased temperature.

Experience tells us that the product $\chi_T \gamma(T,V)/V$ is essentially constant as the temperature varies, which means that, in practice, the linear expansion coefficient is practically proportional to the specific heat capacity at constant volume, i.e.:

$$\alpha \propto C_V \qquad [1.183]$$

Thus, the variation of the expansion coefficient with temperature will be the same in form as that of the specific heat capacity at constant volume, so we obtain the curve shown in Figure 1.16, which has the same shape as Figure 1.14.

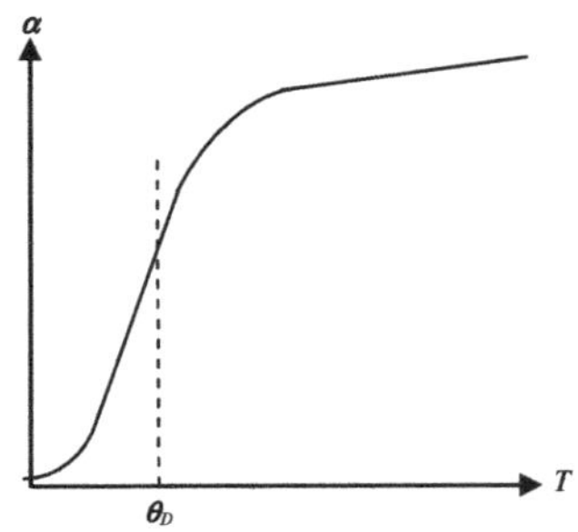

Figure 1.16. *Shape of the curve of the expansion coefficient with temperature in the context of the Debye approximation*

In view of expression [1.183], we can reach the same conclusions as those about the specific heat capacity at constant volume in section 1.9.1.2 (see relations [1.134] and [1.135]):

$$\begin{array}{ll} T \ll \Theta_D & \alpha \propto T^3 \\ T \gg \Theta_D & \alpha \text{ essentially constant} \end{array} \qquad [1.184]$$

Experimentally, the Grüneisen parameter is determined at zero pressure or atmospheric pressure, on the basis of measurements of the volume expansion coefficient β, the adiabatic compressibility (constant entropy) χ_S and the specific heat capacity at constant pressure C_P because we have:

$$\gamma(T,V) = \frac{3\alpha}{\chi_T C_V} = \frac{\beta V}{\chi_S C_P} \qquad [1.185]$$

Relation [1.177] shows that the Grüneisen parameter for the phonon i is dimensionless, and therefore the parameter $\gamma(T,V)$, defined by relation [1.181], is also dimensionless.

In the majority of cases, the Grüneisen parameter is of the order of magnitude of a few units at all temperatures, as is demonstrated by the curves in Figure 1.17, which give the variations of that parameter for several alkali halides.

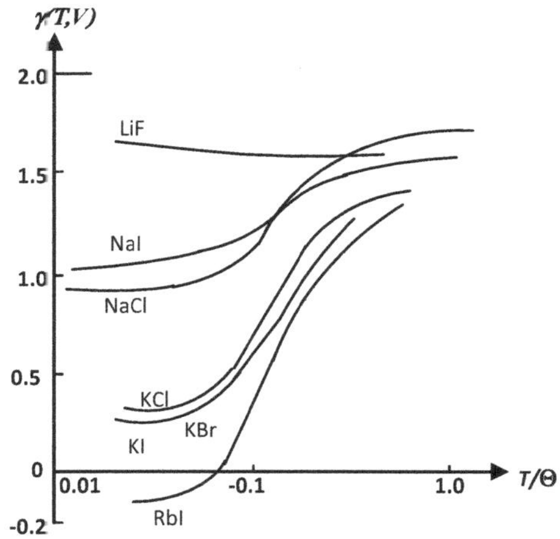

Figure 1.17. *Grüneisen parameter for a few alkali halides (data from [WHI 65])*

However, we note a few cases of values that are far greater – either positive or negative.

The Grüneisen parameter also varies with the volume, as is shown by the example of copper illustrated in Figure 1.18.

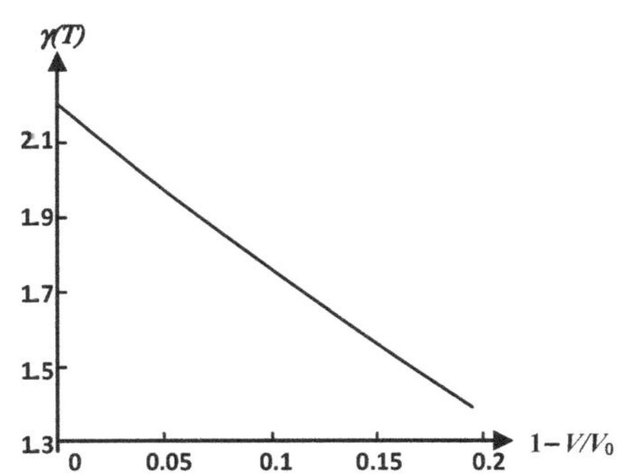

Figure 1.18. *Variation of the Grüneisen parameter for copper with the volume (data taken from [GIR 00])*

1.10.4. *Expansion coefficient of metals*

The reasoning we employed in the previous section is valid for insulating solids. Indeed, for metals, we know that a contribution is made by the free electrons to the partition function and therefore to the specific heat capacity at constant volume. We will now look at the situation for the expansion.

We can think in the case of the electron gas subject to Fermi–Dirac statistics, but calculation shows that we obtain an identical result by using the simple model of the gas of electrons subject to three degrees of freedom of translation, in the case of the equal energy distribution model. If the metal includes n_e electrons per atom, its molar internal energy will be:

$$U_{el} = \frac{3n_e \mathrm{N_a k_B} T}{2} \qquad [1.186]$$

By application of the ideal gas law, we deduce the electron pressure:

$$P_{el} = \frac{n_e \mathrm{N_a k_B} T}{V} = \frac{2U_{el}}{3V} \qquad [1.187]$$

By deriving this pressure in relation to the temperature, we find:

$$\left(\frac{\partial P_{el}}{\partial T}\right)_V = \frac{2}{3V}\left(\frac{\partial U_{el}}{\partial T}\right)_V = \frac{2}{3V} C_{Vel} \qquad [1.188]$$

If we substitute this back into equation [1.170], we find an electron parameter of the Grüneisen constant: $\gamma_{el}(T,V) = 2/3$

Thus, the linear expansion coefficient of the metal will be the sum of two terms: the contribution of the lattice $\gamma_r(T,V)C_{Vr}$ and the contribution of the free electrons $\gamma_{el}(T,V)C_{Vel}$. Hence, we can write:

$$\alpha = \frac{1}{3}\chi_T\left(\gamma_e(T,V)C_{Ve} + \gamma_r(T,V)C_{Vr}\right) \qquad [1.189]$$

At low temperature, we know that the contribution of the lattice varies as the cube of the temperature. By keeping the product $\chi_T \gamma_e(T,V)$ practically constant, the electron contribution also varies as the electron contribution to the specific heat capacity at constant volume. However, as we have seen (relation [1.139]), that contribution varies with T, so, at low temperature, the linear expansion coefficient of a metal will be of the form:

$$\alpha \cong aT + bT^3 \qquad [1.190]$$

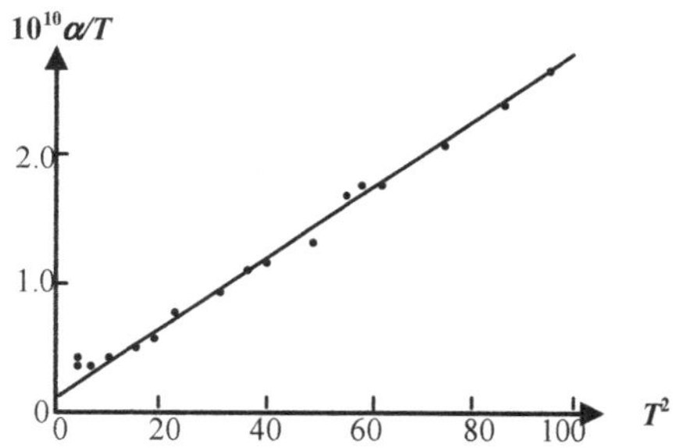

Figure 1.19. *Expansion of copper at low temperature, according to [PER 70]*

Figure 1.19 shows a good example of the application of this law. For copper, it represents the curve showing the α / T ratio as a function of the square of the temperature. This curve is, indeed, a straight line with the equation:

$$\frac{\alpha}{T} \cong 1.3\times10^{-10} + 2.7\times10^{-11}T^2$$

On the basis of this expression, authors have determined the value $\gamma_e(T,V) = 0.57$ for the electron Grüneisen parameter, which is near to the theoretical value of 2/1.

As previously stressed, other properties characterize solids – electrical properties, magnetic properties and other mechanical properties such as resilience, resistance to hardness, etc. All of these properties are not studied here because they are beyond the remit of this book.

2

Solid Solutions

A mixture of two solids A and B may appear in two forms:

– a solid solution – i.e. a single-phase solid with two components A and B;

– a mixture of two phases. These two phases may be two pure phases, respectively, made up of the solids A and B, or a mixture of two solid solutions, each with the two components A and B but with different compositions, or a mixture of one of the two pure solids (for example, A) and a solid solution of A in B.

2.1. Families of solid solutions

The solids A and B may be metal atoms. Mixing them will give us monophasic – or polyphasic alloys or atoms of non-metals. It may also be molecules of organic compounds or polymers. Finally, there are solutions of complex mineral compounds. Thus, alkali feldspars constitute a solid solution between the potassium term orthoclase $(Si_3AlO_8)K$, and the sodium term albite $(Si_3AlO_8)Na$, which is unstable at low temperature.

In certain cases, the solution can have any composition, ranging between pure A and pure B; we, therefore, say that the components A and B are miscible in all proportions. In other cases, the solid A can, at most, only accept a certain proportion of B; in such cases, the limit of solubility of B in A is reached. The lattice of the solid solution is the same as that of the pure solid A. Similarly, the solvent might be B, and may only be able to accommodate, at most, a certain proportion of A, which is then the limit of solubility of A in B. In this latter case, it is the lattice of the pure solid B which is the lattice of the solution. The solid solutions thus obtained are known as primary solid solutions.

It occurs that the two components A and B can react with each other to give a chemical compound which we call a definite compound, with the formula A_aB_b. In general, this compound has its own crystalline lattice, which is different from those of the pure substances A and B. This definite compound can react with A or B to give solid solutions which, finally, constitute a new type of solid solution, formed of the components A and B. These new solutions are called secondary solutions.

We usually distinguish two types of solid solutions.

2.1.1. *Substitutional solid solutions*

We distinguish two types of solids: those in which the nodes of the lattice are all identical, such as metals or, more generally, solids with covalent bonds and solids with two families of lattice nodes, such as ionic compounds.

In solids containing only one family of nodes, the atoms (or molecules) of the two components are placed on the nodes of the single lattice, the coordination index is the same for both components. For instance, with a binary solution of the solvent A and solute B, the crystalline lattice is that of the pure A and, at certain sites, atoms of B substitute atoms of A in the initial structure, but without modifying that structure.

The limit of solubility of B in A, to form a substitutional solution, is governed – particularly with metal alloys – by the empirical rules devised by Hume and Rothery. According to these rules, there are four factors which determine the degree of solubility of a substance B in a substance A:

– *structural rule*: the crystallographic structures of the solids A and B must match;

– *electron structural rule*: electronically similar metals can form extended solutions because of the similarity of their bond;

– *valence rule*: if they have the same valence, the metals A and B will dissolve one another easily. However, metals with a low valence will more easily dissolve a metal with a similar valence than metals with a high valence;

– *atomic radius rule* (or atomic diameter rule, or "15% rule"): in order for complete miscibility to occur, the difference between the atomic radii must be no greater than 15%, meaning that we must have:

$$\frac{2\left|r_A - r_B\right|}{r_A + r_B} \leq 0.15 \qquad [2.1]$$

In the case of non-total miscibility, the solubility is inversely proportional to the difference in size of the atoms.

Let us take the example of copper–nickel alloys. Table 2.1 lists some of the characteristics of the atoms. We can see that the two pure metals have the same structure, have extremely similar atomic radii and electronegativity values, on the Pauling scale, which are also very close. Copper and nickel are indeed miscible in all proportions.

Metals	Crystalline lattice	Atomic radius (nm)	Electronegativity
Copper	Ccf	0.128	1.9
Nickel	Ccf	0.125	1.8
Zinc	Hc	0.133	1.6

Table 2.1. *Comparison of the characteristics of copper and nickel*

As another example, let us take a look at copper–zinc alloys. Table 2.1 shows quite different characteristics of the two elements; in particular, they crystallize in two different crystalline systems (centered cubic faces for copper and hexagonal compact for zinc). Although the 15% rule is respected (there is a difference of 4%), these two metals are not miscible in all proportions, which shows that this rule is not sufficient, alone, to ensure total solubility. In addition, we can see that zinc is more soluble in copper than *vice versa*. This is attributable to the fact that the valences of the two metals are different – the valence of zinc is 2, whereas that of copper is 1.

In the case of ionic compounds, the two families of nodes of the lattice each form a sublattice. Generally, when the substitution is made, the nature of the substituent substance is respected. Thus, cations can only be substituted by cations and anions can only be substituted by anions. For example, sodium chloride and potassium chloride are miscible in all proportions. The sodium and potassium ions are interchangeable in those substances.

Mineral solids such as the feldspars mentioned above are often the site of such substitutions.

The solubility limit is reach even more quickly when the substituted ions are different from the substituting ions. Solubility is favored by crystallization in an identical crystalline system in the two pure solids, the same charge on the ions and similar ionic radii.

If the substituting ion has a different charge from that of the substituted ion, the conservation of electrical neutrality causes modifications, which will be studied in Chapter 3.

In numerous cases, known as reciprocal solutions, the solid solution can be considered to be the product of two pairs of pure compounds, which are called the poles. Thus, the solution between iron sesquioxide (Fe_3O_4) and nickel chromate ($NiCr_2O_4$) yields spinel $(Fe^{2+}, Ni^{2+})_{1-x}(Cr^{3+}, Fe^{3+})_x$. However, this solution could also give rise to the following pure compounds: iron chromate ($FeCr_2O_4$) and nickel ferrite ($NiFeO_4$). We say that the possible poles of spinel are either the iron sesquioxide–nickel chromate pair or the iron chromate–nickel ferrite pair.

The density of the binary substitutional solution of A and B is calculated by the relation:

$$\rho = \frac{n\left(x_{\mathrm{A}} M_{\mathrm{A}} + x_{\mathrm{B}} M_{\mathrm{B}}\right)}{V \mathrm{N}_{\mathrm{a}}} \qquad [2.2]$$

In this relation, n is the number of atoms per mesh, V is the volume of the mesh, N_a is Avogadro's constant, x_A and x_B are the molar fractions of each of the components, whose, respectively, molar masses are M_A and M_B.

When there are a higher number of components, relation [2.2] can be generalized, and the density is:

$$\rho = \frac{n\sum_i x_i M_i}{V\mathrm{N_a}} \qquad [2.3]$$

2.1.2. *Insertion solid solution*

The atom (or molecule) of the component of the smallest volume is placed in the interstices of the atoms (or molecules) of the other component. The coordination indices of the two species are no longer necessarily equal.

In the context of compact lattices: cubic centered faces (ccf), cubic centered (cc) and hexagonal compact (hc), there are two types of insertion sites: octahedral and tetrahedral sites.

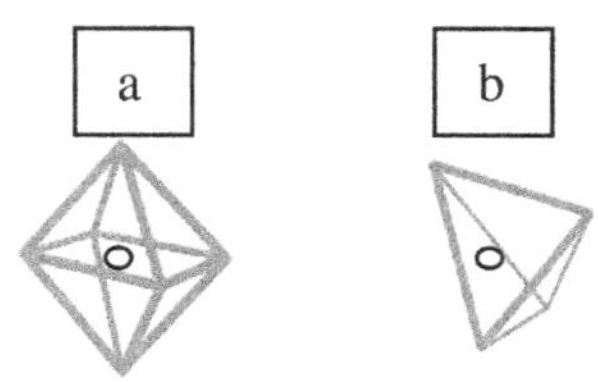

Figure 2.1. *Insertion sites in a compact lattice: a) octahedral site and b) tetrahedral site*

Octahedral sites (Figure 2.1(a)) are characterized by six equidistant near neighbors which form an octahedron, while tetrahedral sites (Figure 2.1(b)) contain only four equidistant near neighbors, forming a tetrahedron. In principle, it is the site whose dimension is nearest to the radius of the inserted atom which is occupied. Maximum solubility is obtained when all the interstitial sites of the same nature are occupied.

The inserted atoms are usually those which have the smallest atomic radius – i.e. essentially atoms of non-metals at the top of the periodic classification. Table 2.2 shows the atomic radii of these atoms which are inserted most often.

Atom	H	B	C	N	O
Atomic radius (nm)	4.6	9.7	7.7	7.1	6.0

Table 2.2. *Atomic radii of the atoms that are most frequently inserted into a solid solution*

Let us look at the position, saturation and dimension of the insertion sites of compact stacks.

2.1.2.1. *Octahedral sites of the cubic centered faces lattice*

In the cubic centered faces system, we find (Figure 2.2(a)) an octahedral interstitial site at the center of the mesh of coordinates (1/2, 1/2, 1/2) and a site at the middle of each edge (1/2, 0.0). Thus, we find four sites per mesh. As the system contains four atoms per mesh, we have one interstitial site per atom. Thus, saturation is attained for a molar fraction of solute: $x_B = 0.5$.

In Figure 2.2(b), we easily calculate the relation between the different values: the mesh parameter a, the atomic radius of the solvent r_A and the radius of the insertion site – i.e. the radius r_i of the largest sphere which it is possible to insert:

$$r_A = \frac{\sqrt{2}a}{4} \qquad [2.4]$$

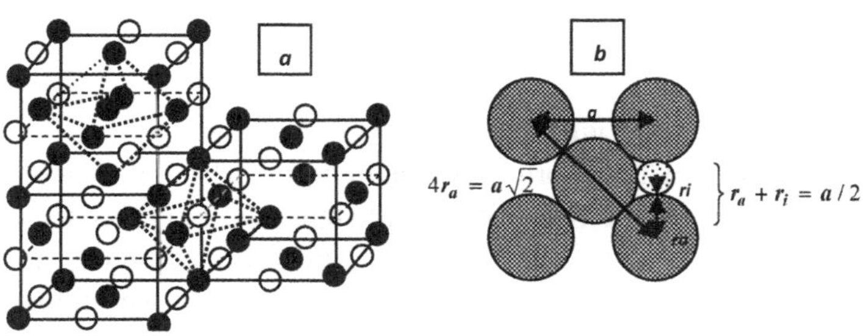

Figure 2.2. *Octahedral sites in a cubic lattice with centered faces*

$$r_i = 0.47a \qquad [2.5]$$

$$r_i = r_A\left(\sqrt{2}-1\right) = 0.414 r_A \qquad [2.6]$$

If we look at the example of the insertion into austenite (γiron), the mesh parameter there is $a = 35.4$ nm. If we compare the dimension of the corresponding octahedral site ($r_i = 5.2$ nm), calculated using relation [2.5], with the atomic radii of the atoms that are most frequently inserted (Table 2.2), we see that only the hydrogen atom has an atomic radius that is smaller than the insertion radius, and therefore in order for other atoms to be inserted into an octahedral site, the lattice would need to be expanded.

2.1.2.2. *Tetrahedral sites of the cubic centered faces lattice*

In the cubic centered faces system, we find (Figure 2.3(a)) a tetrahedral interstitial site a quarter of the way along the major diagonals, with coordinates (1/4, 1/4, 1/4). Thus, we have eight sites per mesh. As the system contains four atoms per mesh, we have one interstitial site for two atoms. Hence, saturation is attained for a molar fraction of solute: $x_B = 0.33$.

In Figure 2.2(b), we easily calculate the relation between the different values: the mesh parameter a, the atomic radius of the solvent r_A and the radius of the insertion site – i.e. the radius r_i of the largest sphere which it is possible to insert.

We obtain:

$$r_i = 0.079a \qquad [2.7]$$

$$r_i = r_A\left(\frac{\sqrt{3}}{\sqrt{2}} - 1\right) = 0.225 r_A \qquad [2.8]$$

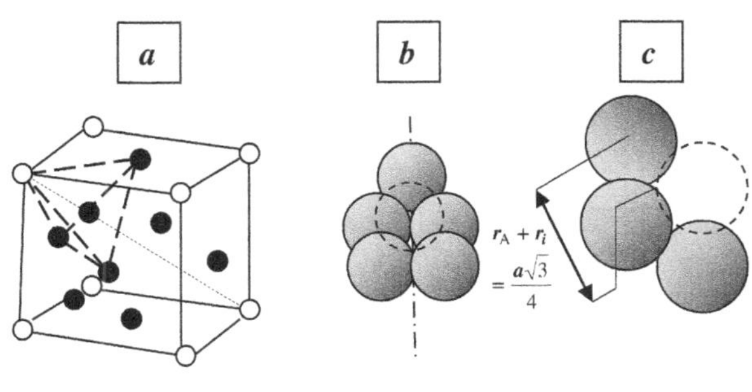

Figure 2.3. *Tetrahedral sites in a cubic system with centered faces*

Looking again at the example of insertion into austenite (γ iron), the mesh parameter is $a = 35.4$ nm. If we compare the dimension of the corresponding tetrahedral site ($r_i = 2.8$ nm), calculated using relation [2.7], with the atomic radii of the atoms most frequently inserted (Table 2.2), we can see that the insertion of any one of these atoms would require the expansion of the lattice.

2.1.2.3. *Octahedral sites in the cubic centered lattice*

In the cubic centered system, we find (Figure 2.4(a)) an octahedral interstitial site at the center of each face, with coordinates (1/2, 1/2, 0). Therefore, we count six sites per mesh. As the system contains two atoms per mesh, therefore, we have three interstitial sites per atom. Hence, saturation is attained for a molar fraction of solute: $x_B = 0.75$.

Using the values defined in Figure 2.4(b), we can easily calculate the characteristic values.

$$r_h = 0.274a \quad [2.9]$$

$$r_h = 0.632 r_A \quad [2.10]$$

$$r_i = r_k = 0.067a \quad [2.11]$$

$$r_i = r_k = \left(\frac{2}{\sqrt{3}} - 1\right) r_A = 0.154 r_A \quad [2.12]$$

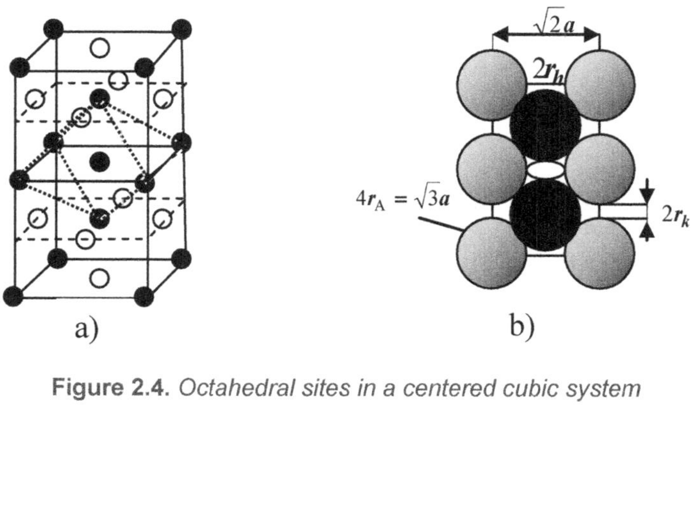

Figure 2.4. *Octahedral sites in a centered cubic system*

If we take the example of insertion into ferrite (α iron), the mesh parameter is a = 28.6 nm. If we compare the radius of the corresponding octahedral site (r_i = 1.9 nm), calculated using relation [2.11], with the atomic radii of the atoms which are most commonly inserted (see Table 2.2), we see that all the atoms in the table can only be inserted if the lattice is expanded.

2.1.2.4. *Tetrahedral sites in the cubic centered lattice*

In the cubic centered system, we find (Figure 2.5(a)) a tetrahedral interstitial site at the quarter-points of the edges and halfway along the sides, with coordinates (1/2, 1/4, 0). Therefore, we have 12 sites per mesh. As the system contains two atoms per mesh, therefore, we have six interstitial sites per atom. Thus, saturation is attained for a molar fraction of solute: $x_B = 6/7 = 0.85$.

With the values defined in Figure 2.5(b), we can easily calculate the characteristic values.

$$r_i = 0.126a \quad [2.13]$$

$$r_i = 0.291 r_A \quad [2.14]$$

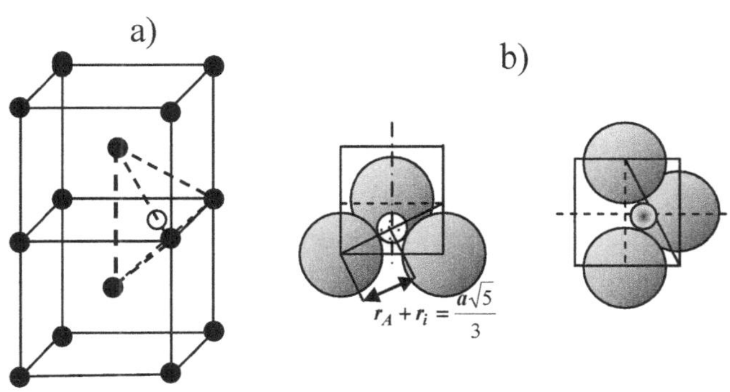

Figure 2.5. *Tetrahedral sites in a centered cubic system*

If we look again at the example of insertion into ferrite (α iron), the mesh parameter is a = 28.6 nm. Comparing the radius of the corresponding octahedral site (r_i = 3.6 nm), calculated on the basis of relation [2.13], with the atomic radii of the atoms that are most frequently inserted (Table 2.2), we can see that all of the atoms in the table can only be inserted if the lattice is expanded.

2.1.2.5. *Octahedral sites in the hexagonal compact lattice*

In the hexagonal compact system, we find (Figure 2.6), as for the cubic centered face lattice, four octahedral interstitial sites per mesh, and hence one interstitial site per atom. Indeed, the two structures hexagonal compact (HC) and CFC are both compact stacks of compact planes. The only difference between the two is the relative position of the third plane.

Thus, saturation is attained for a molar fraction of solute: $x_B = 0.5$.

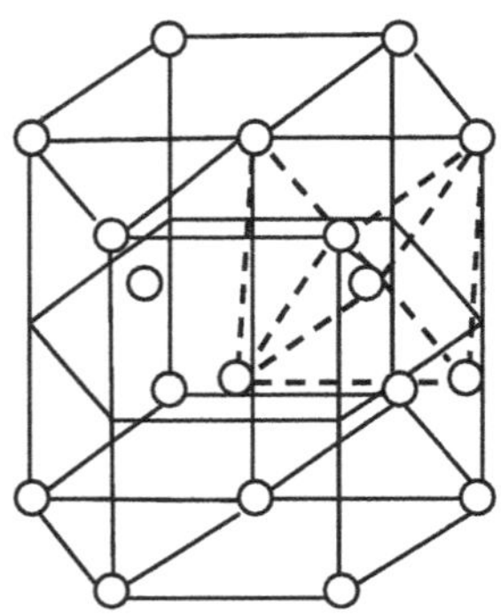

Figure 2.6. *Octahedral sites in a hexagonal compact structure*

We can use relations [2.4]–[2.6] to determine the insertion radius. The insertion conditions are, therefore, the same as in cubic centered face lattices.

2.1.2.6. *Tetrahedral sites in the hexagonal compact lattice*

Figure 2.7 shows the position of a tetrahedral site in a hexagonal compact lattice. As in the previous case, the conditions of insertion into the hexagonal compact lattice will be the same as in the tetrahedral sites of the cubic centered face lattice – i.e. one interstitial site for two atoms, which means saturation $x_B = 0.33$. Relations [2.7] and [2.8] remain valid for the ratios between the characteristic values.

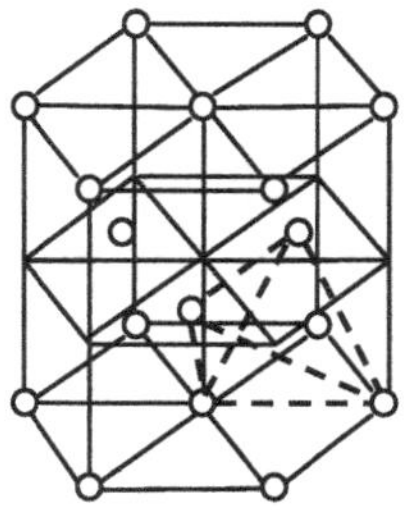

Figure 2.7. *Tetrahedral sites in a hexagonal compact lattice*

2.1.2.7. *Condition of insertion in the different compact lattices*

By comparing relations [2.6] and [2.8] for the cubic centered face system and the hexagonal compact system, we can see that the octahedral site is larger than the tetrahedral site. Therefore, insertion in such lattices will tend to take place more on octahedral sites. The insertion radius will, therefore, be $r_i = 0.414 r_A$.

By comparing relations [2.12] and [2.14], for the cubic centered face system, we see that the octahedral site is smaller than the tetrahedral site. Hence, insertion in such lattices will tend to take place on tetrahedral sites. The insertion radius will, therefore, be $r_i = 0.291 r_A$.

2.1.2.8. *Density of an insertion solution*

We can easily calculate that, in a binary insertion solution, the density is given as a function of the composition x_A of the solution by the relation:

$$\rho(\mathrm{A}) = \frac{n\left(x_A M_A + x_B n_B\right)}{\mathrm{N_a} V x_A} \qquad [2.15]$$

For a solution with more than two components, we also calculate the density:

$$\rho(\mathrm{A}) = \frac{n\sum_i x_i M_i}{V\mathrm{N_a}\sum_{j\neq \mathrm{A}} x_j} \qquad [2.16]$$

We have only looked at the case where the lattice was made up of all identical sites – i.e. sites which had the same coordination index. There is another kind of solid solution in which the coordination indices of the two components are not necessarily equal – that in which the atoms of each component are placed at the nodes of a distinct sublattice, so the lattice is formed of two families of non-identical sites. This type of solution often gives us definite compounds, which frequently exhibit major deviations from stoichiometry (see Chapter 3).

2.2. Order in solid solutions

In the case of a substitutional solid solution, the atoms of the solute B can occupy a variety of positions in relation to the atoms of the solvent A.

The atoms of solvent and the atoms of solute may be distributed utterly randomly among the various sites of the lattice, in which case the solution is said to be "disordered". Such is the case when the two elementary components are perfectly equivalent (with the same crystalline structure and similar atomic dimensions). Figure 2.8(a) offers an illustration of such an entirely disordered solution. The gray circles and the black circles, which each represent one type of component, are distributed completely at random.

In other cases, there is a more or less marked tendency toward the acquisition of an order. This might be an ordered position of the solute in relation to the solvent, as in the example of an atom of solvent which tends to be surrounded by atoms of solute. We then have a "short-distance order". It might also be a systematic position of the solute atoms in relation to the solvent atoms, with a spatial periodicity to that arrangement. In this case, we have a *long-distance order*, as represented in Figure 2.8(b), which shows a periodic distribution of the gray circles and a periodic distribution of the black circles. We then say that we have an *ordered* solution. It is also possible to find collections of atoms of the same type, leading to *clusters*– i.e. areas of space which appear to be single-component phases. Figure 2.8(c) gives a diagrammatic representation of this kind of arrangement. There are zones which are completely filled with gray circles, and others which are entirely filled with black circles. This ordering often leads to demixing of the solid solution into two distinct phases.

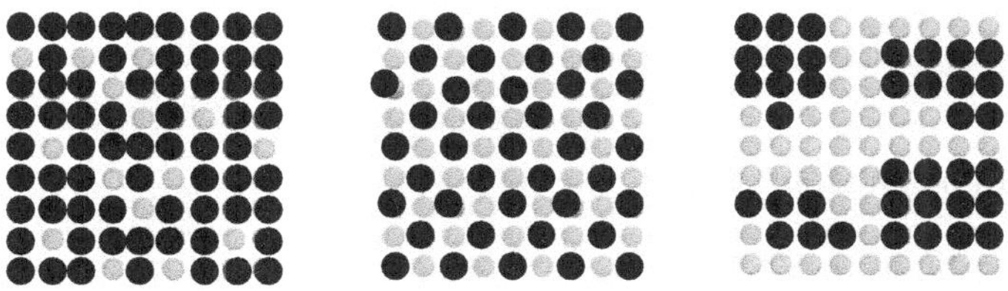

a) b) c)

Figure 2.8. *Diagrammatic representation: a) completely disordered solution; b) ordered solution and; c) solution with single-component clusters*

2.2.1. *Short-distance order*

We find the concept of local composition in solid solutions, just as it is found for liquid solutions. Because of the existence of an exchange energy, we can imagine that if a molecule of A and a molecule of B experience a greater force of attraction than two molecules of A or two molecules of B, a molecule of A will tend to be surrounded by molecules of B and therefore the composition of the immediate environment of the molecule of A will not be the same as the overall composition of the solution: the immediate vicinity is richer in molecules of B. The opposite effect is obtained if the molecules A and B attract each other less strongly than do two molecules of A or two molecules of B. This is the *concept of local composition*, which was first put forward by Wilson.

To describe this concept of local composition, we use a nomenclature which enables us to clearly distinguish, every time, between the central molecule and its nearest surrounding molecules.

We use the notation N_{ij} to speak of the number of molecules of i immediately neighboring a molecule of j. The probability of finding a molecule of i immediately neighboring a molecule of j is given by the following relation (where N denotes the number of components in the solution):

$$P_{ij} = \frac{N_{ij}}{\sum_{i=1}^{N} N_{ij}} \quad [2.17]$$

This number also represents the local molar fraction of the atoms of i around an atom of j.

Clearly, we have:

$$\sum_{i=1}^{N} P_{ij} = 1 \quad [2.18]$$

Thus, for a binary solution, we distinguish:

– P_{AA}: local molar fraction of the molecules of A around a molecule of A;

– P_{BA}: local molar fraction of the molecules of B around a molecule of A;

– P_{BB}: local molar fraction of the molecules of B around a molecule of B;

– P_{AB}: local molar fraction of the molecules of A around a molecule of B.

with the equalities:

$$P_{AA} + P_{BA} = 1 \quad [2.19a]$$

$$P_{BB} + P_{AB} = 1 \quad [2.19b]$$

For a solution containing molecules i and j, Warren and Cowley defined an order parameter η_{ij} as:

$$\eta_{ij} = 1 - \frac{\text{number of molecules of } i \text{ around a molecule of } j}{\text{average number of molecules of } i \text{ around a molecule of } j} \quad [2.20]$$

Thus, in the case of a binary mixture, we would have:

$$\eta_{\mathrm{AB}} = 1 - \frac{P_{\mathrm{AB}}}{x_{\mathrm{A}}} \qquad [2.21]$$

with the condition: $0 < \eta_{\mathrm{AB}} \leq 1$ [2.22]

and symmetrically:

$$\eta_{\mathrm{BA}} = 1 - \frac{P_{\mathrm{BA}}}{x_{\mathrm{B}}} \qquad [2.23]$$

and the condition $0 < \eta_{\mathrm{BA}} \leq 1$ [2.24]

Distinctly, we can deduce from this that:

$$P_{\mathrm{BA}} = x_{\mathrm{B}}\left(1 - \eta_{\mathrm{BA}}\right) \qquad [2.25a]$$

$$P_{\mathrm{AB}} = x_{\mathrm{A}}\left(1 - \eta_{\mathrm{AB}}\right) \qquad [2.25b]$$

NOTE.– If the molecules are randomly distributed, then $P_{\mathrm{AB}} = x_{\mathrm{A}}$ and $\eta_{\mathrm{BA}} = 1$.

The two order parameters η_{AB} and η_{BA} are interlinked – indeed, the number of pairs A-B can be expressed in two different ways, such that:

$$N_{\mathrm{BA}} N_{\mathrm{A}} = N_{\mathrm{AB}} N_{\mathrm{B}} \qquad [2.26]$$

Thus:

$$z'_{\mathrm{A}} P_{\mathrm{BA}} N_{\mathrm{A}} = z'_{\mathrm{B}} P_{\mathrm{AB}} N_{\mathrm{B}} \qquad [2.27]$$

By introducing the values of P_{BA} and P_{AB} from equations [2.25a] and [2.25b], we obtain:

$$z'_{\mathrm{A}} x_{\mathrm{B}} N_{\mathrm{A}} (1 - \eta_{\mathrm{BA}}) = z'_{\mathrm{B}} x_{\mathrm{A}} N_{\mathrm{B}} (1 - \eta_{\mathrm{AB}}) \qquad [2.28]$$

Noting that we have the equality:

$$x_{B}N_{A} = x_{A}N_{B} = \frac{N_{A}N_{B}}{N_{A} + N_{B}} \tag{2.29}$$

we find:

$$z'_{A}\eta_{BA} - z'_{B}\eta_{AB} = z'_{B} - z'_{B} \tag{2.30}$$

Hence, the two parameters of opposite orders are not independent.

In a substitutional solid solution with a single type of site, the coordination indices of the two types of molecules are the same in the solution ($z'_{A} = z'_{B}$), so the two Warren and Cowley order parameters are also equal ($\eta_{AB} = \eta_{BA}$).

In view of the relations [2.25a], if $\eta_{AB} = 0$, then the distribution of molecules A and B is random.

If $\eta_{AB} > 0$, this means that $P_{AB} < x_{A}$. Hence, on average, there are fewer mixed pairs than there are in the disordered solution, which leads to a tendency toward separation of the entities A and B (demixing). However, if $\eta_{AB} < 0$, this would mean that there is a certain tendency toward the association of the molecules of A with the molecules of B, so solution tends to become ordered.

As the temperature increases, the short-distance order decreases, because of two mutually complementary phenomena: the exchange energies decrease and the thermal agitation increases. The solution then tends toward random distribution.

When we decrease the temperature, the exchange energies increase and, below a certain critical temperature T_c, the interactions begin to be felt at a long distance and we obtain a long-distance order (see section 2.2.2).

This notion of short-distance order comes into play in certain solution models, and is encountered in the quasi-chemical model (see section 2.3.4).

2.2.2. *Long-distance order*

The concept of long-distance order introduced previously is, practically, encountered only for metal alloys. For this reason, we limit our remarks here to solutions of atoms.

When we decrease the temperature of a solid solution of metal atoms, we see the emergence of distinct families of sites which form sublattices. Thus, each sublattice is occupied by a specific type of atom. When each sublattice is occupied by a single type of atom, the solution is completely ordered: this is "long-distance order". Ordering is generally accompanied by a drop in symmetry.

2.2.2.1. *Order and disorder in an alloy*

Without going into crystallographic detail about the different types of ordered solutions, let us give a few simple examples.

Figure 2.9 shows the ordering of copper and gold in the equi-atomic alloy Au-Cu. In the disordered phase, all the sites of a cubic centered faced crystal are randomly occupied by atoms of copper or gold. The probability that a given site will be occupied by an atom of copper is ½. The same is true of the probability of a site being occupied by an atom of gold. This is represented by the presence, at each site, of a semi-shaded circle.

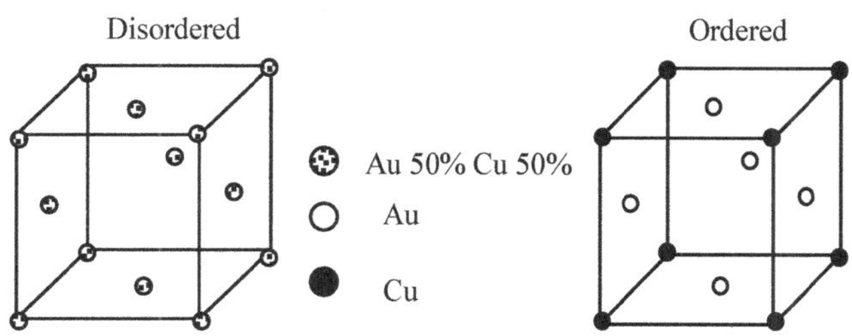

Figure 2.9. *Order in the alloy CuAu*

For the ordered phase, if the copper atoms occupy the vertices of the previous cube (black circles), then the gold atoms are situated on the faces of the cubes (white circles). Thus, the crystal exhibits two cubic sublattices.

In our example, all the crystalline sites initially play the same role in the disordered phase: it is only when the phase is ordered that we can distinguish two types of sites. The notion of distinct sites thus occurs in the ordered phase of the alloy.

Let us now examine the example of alloys with the composition Fe_3Al (Figure 2.10).

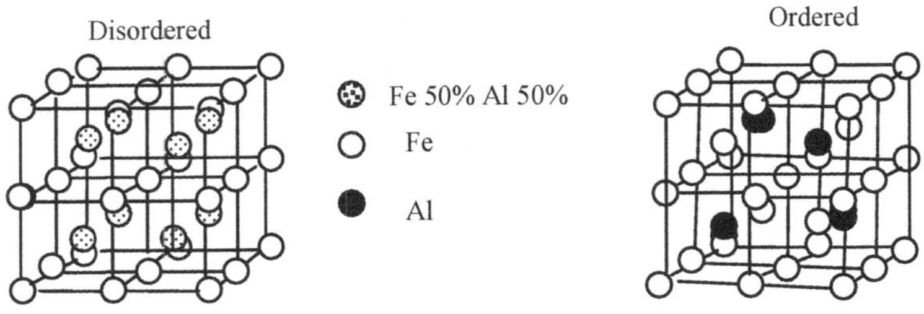

Figure 2.10. *Order in the alloy Fe_3Al*

The disordered phase exhibits two types of sites: one which is always occupied by iron atoms and second which is randomly occupied by either an iron atom or an aluminum atom, with a 50% probability of finding one element or the other. Thus, the iron atoms are not all placed equally in the disordered crystal. The two tiers of iron atoms occupy a simple cubic crystal in all conditions. The remaining iron atoms, along with the aluminum atoms, randomly occupy the vertices of a simple cubic crystal.

In the ordered solution, the iron sites of the first species always occupy their specific sites, while atoms are ordered around the other sites, one out of every two sites is occupied by an iron atom and one out of two sites is occupied by an aluminum atom. One-third of the iron atoms and the aluminum atoms form a cubic crystal with centered faces such as NaCl. Thus, in this example, not all atoms are affected by the order-disorder arrangement. In the completely disordered solution, there are two types of sites: those which are always occupied by iron and those which are randomly occupied by either iron or aluminum. In the ordered solution, again we have two types of sites: those occupied by iron and those occupied by aluminum. In fact, because of their origin, we can clearly see that the sites occupied by iron in the ordered phase must be distinguished and thus there must be three types of sites in the ordered phase.

It is clear that, in order for such exchanges of atoms between sites to be possible, it is necessary for these atoms to have relatively similar dimensions. Any difference in volume necessarily leads to the creation of stresses which the solid must be able to withstand.

On an experimental level, order in a solid solution can be detected by X-ray diffraction (see section 2.4.5).

2.2.2.2. *Degree of long-distance order*

We have perfectly defined the two states of order and disorder. However, we can imagine intermediary states – e.g. a state where a certain number of atoms of A are on sites that are normally attributed to A in the perfectly ordered solution. This number would obviously be between the average and the total number of atoms of A. In order to characterize such an intermediary state, Bragg and Williams defined a *degree of order* or *long-distance order parameter s*, such that this degree is equal to 1 if the solution is perfectly ordered, and 0 in the case of a completely random distribution solution.

With this in mind, we start with the completely ordered state and use the notation A_A to denote an atom of A placed on a site attributed to A and B_B for an atom of B placed on a site attributed to B. If the order is not complete, we would have four types of atoms:

– atoms A_A: atoms of A placed on sites attributed to A in the ordered solution; their number is $N_{A(A)}$;

– atoms B_B: atoms of B placed on sites attributed to B in the ordered solution; their number is $N_{B(B)}$;

– atoms A_B: atoms of A placed on sites attributed to B in the ordered solution; their number is $N_{A(B)}$;

– B_A: atoms of B placed on sites attributed to A in the ordered solution; their number is $N_{B(A)}$.

Let us use the symbol x to represent the molar fraction of the element A in the solid solution; obviously, the molar fraction of the element B will be $1-x$.

We will use y to denote the fraction of sites normally occupied by the atom A in the perfectly ordered alloy. The fraction of sites occupied by B will be $1-y$.

NOTE.– In the definition of the fraction of sites, in the denominator, we see the sum of all the sites on the lattice. It is clear that this sum only takes account of the sites which are "active" in the order/disorder operation. For example, in the case of the solution Fe_3Al we looked at previously, only the sites shown in gray in Figure 2.10 in the disordered alloy will be counted. The other sites in the iron, which are inactive, are not taken into account.

We suppose that the lattice contains no vacancies, meaning that each and every node is occupied either by an atom A or by an atom B. This means that the total number of sites $N_{(A)}+N_{(B)}$ is equal to the total number of atoms N_A+N_B, so:

$$N = N_A + N_B = N_{(A)} + N_{(B)} \qquad [2.31]$$

In given temperature conditions, the solid solution will contain:

– N_{A_A} atoms of A placed on sites of A;

– N_{A_B} atoms of A placed on sites of B;

– N_{B_A} atoms of B placed on sites of A;

– N_{B_B} atoms of B placed on sites of B.

The molar fraction of A and the fraction of sites of type A, are, therefore:

$$x = \frac{N_A}{N_A + N_B} \qquad [2.32a]$$

$$y = \frac{N_{(A)}}{N_{(A)} + N_{(B)}} \quad [2.32b]$$

The probabilities of occupation of the sites of A and B by the two types of atoms will, therefore, be given by:

$$P_{A_A} = \frac{N_{A_A}}{N_{(A)}} \quad [2.33a]$$

$$P_{A_B} = \frac{N_{A_B}}{N_{(B)}} \quad [2.33b]$$

$$P_{B_A} = \frac{N_{B_A}}{N_{(A)}} \quad [2.33c]$$

$$P_{B_B} = \frac{N_{B_B}}{N_{(B)}} \quad [2.33d]$$

All these probabilities are linked to each other by simple relations:

$$P_{A_A} + P_{B_A} = 1 \quad [2.34a]$$

$$P_{A_B} + P_{B_B} = 1 \quad [2.34b]$$

We can also write, on the basis of the definitions of the molar fractions and the fractions of sites:

$$x = yP_{A_A} + (1 - y)P_{A_B} \quad [2.35]$$

$$1 - x = (1 - y)P_{B_B} + yP_{B_A} \quad [2.36]$$

The degree of order at long distance, *s*, or the Bragg and Williams degree of order[1], is defined by the relation:

1 Other authors define the degree of order differently. For example, Borelius defines a degree of order which corresponds to $1\text{-}s^2$.

$$s = P_{A_A} - P_{A_B} = P_{B_B} - P_{B_A} \qquad [2.37]$$

x, y	N_{A_A}	N_{A_B}	N_{B_A}	N_{B_B}
x, y	$N_{(A)}(x+ys)$	$N_{(B)}(x-ys)$	$N_{(A)}[1-x-ys]$	$N_{(B)}[1-x+ys]$
$x=y$	$N_{(A)}x(1+s)$	$N_{(B)}x(1-s)$	$N_{(A)}[1-x(1-s)]$	$N_{(B)}[1-x(1+s)]$
$y = ½$	$\frac{N}{2}\left(x+\frac{s}{2}\right)$	$\frac{N}{2}\left(x-\frac{s}{2}\right)$	$\frac{N}{2}\left(1-x-\frac{s}{2}\right)$	$\frac{N}{2}\left(1-x+\frac{s}{2}\right)$
$x=y=1/2$	$\frac{N}{4}(1+s)$	$\frac{N}{4}(1-s)$	$\frac{N}{4}(1-s)$	$\frac{N}{4}(1+s)$

Table 2.3. *Populations of the different sites of an alloy as a function of the degree of order (for any given values of x and y, for x = y, for y = ½ and for x = y = ½)*

Thus, by combining relations [2.32]–[2.35], the populations of the different sites are calculated as a function of this parameter s. The expressions obtained are given in the first row of Table 2.3.

We will now examine the values of the degree of order in the extreme cases.

If the alloy is completely disordered, then the probability of finding an atom A on a site of A is given by the molar fraction of the atoms of A – i.e. x – and therefore the order parameter is null: $s = 0$.

If the alloy is ordered as fully as possible, two scenarios can be envisaged:

1) If the alloy's composition is stoichiometric – i.e. if the atomic fraction of A is identical to the fraction of sites A ($x = y$) – then the degree of order reaches its maximum value ($s = 1$). The quantities of the different combinations are shown in the second row of Table 2.3.

2) If the alloy's composition is not stoichiometric ($x \neq y$), then even in an ordered alloy, certain atoms are not on their usual sites.

If $x > y$, all the sites of A are occupied by atoms of A, but as they are more numerous, the remaining atoms of A will be placed on sites usually reserved for B, and thus we have:

$$N_{A_A} = N_{(A)} \quad [2.38a]$$

and

$$N_{A_B} \neq 0 \quad [2.38b]$$

In this case, the degree of order will reach a maximum value given by:

$$s_{max} = \frac{1-x}{1-y} \quad [2.39]$$

We can see that this maximum value is less than 1.

If $x < y$, all the sites of B are occupied by atoms of B, but as there are more of them, the other atoms of B are placed on the sites for A, and thus we have:

$$N_{A_A} = \frac{x}{y} \quad [2.40]$$

In this case, the degree of order will reach a maximum value given by:

$$s_{max} = \frac{x}{y} \quad [2.41]$$

We can see that this maximum value is still less than 1.

Hence, the value of the degree of order will, in all cases, be between zero, which corresponds to maximum disorder, and a maximum value, which can reach up to 1 in the case of stoichiometric alloys, where the order is perfect.

2.3. Thermodynamic models of solid solutions

The models used to describe the thermodynamic properties of solid solutions are the same as for liquid solutions. The hypotheses upon which these models are founded can be easily applied – all the more because the hypothesis of the pseudo-lattice used for liquid solutions now becomes perfectly appropriate, and is no longer a hypothesis, but simply corresponds to the crystalline lattice.

In these models, we are merely interested in obtaining the mixing values in order to find the Gibbs energy of mixing, and therefore the activities. Thus, we are not concerned with the models of solids used for the pure components. Thus, the models giving the mixing values are, therefore, independent of the solid-state models chosen, which we saw in Chapter 1.

Essentially, we are looking at the solubility of metals in other metals – i.e. monophase metal alloys. The most commonly used solution models are the models with similar atomic volumes, which give us the perfect solution, the infinitely-dilute solution and the strictly-regular solution. Thus, we will then look at Guggenheim's quasi-chemical model, which includes the notion of short-distance order.

2.3.1. *Determination of the Gibbs energy of mixing*

Let us consider a solid binary solution: a mixture of the two components A and B, which contains n_A moles of component A and n_B moles of component B – i.e. N_A molecules of A and N_B molecules of B. This solution satisfies the following five conditions:

1) The crystalline lattice of the solid solution is very similar to the lattices of each of the pure solids A and B. In particular, these three solids exhibit the same coordination index z (say, 12 for the hexagonal lattice or the cubic, centered face lattice).

2) The molar volumes v_A^0 and v_B^0 of the two pure solids are sufficiently similar. Certain authors have evaluated this condition to be acceptable if the ratio between the radii of the two molecules A and

B (which are supposed to be spherical) is no greater than 1.25. This property is very commonly found with monophase metal alloys.

3) The mixing volume is null, which means that the volume of the solution is given by the additive law:

$$V = n_{\mathrm{A}} v_{\mathrm{A}}^{0} + n_{\mathrm{B}} v_{\mathrm{B}}^{0} \quad [2.42]$$

4) The potential energy of interaction E_I can be considered to be the sum of the contributions of the most closely neighboring pairs of molecules.

5) $\varepsilon_{\mathrm{A(A)}}$ and $\varepsilon_{\mathrm{B(B)}}$, respectively, denote the energy necessary to extract an atom of A or B from the solution and convert it into a perfect gaseous phase where we consider that the potential energy of each atom is null (lattice energies). Thus, the potential energy of an atom A measured with reference to the monatomic perfect gas A is $-\varepsilon_{\mathrm{A(A)}}$. The minus sign simply indicates that in the chosen temperature conditions, the solid state is more stable than the gas. Similarly, we would have the potential energy $-\varepsilon_{\mathrm{B(B)}}$ for the atom B in relation to the pure perfect gas B.

If we consider a pair of close neighbors A-A, each atom A contributes for a part of $\varepsilon_{\mathrm{A(A)}} / z$ to the energy of the pair A-A.

The average energy of interaction between two molecules of A is, therefore, $-2\varepsilon_{\mathrm{A(A)}} / z$, and similarly for two molecules of Bit is $-2\varepsilon_{\mathrm{B(B)}} / z$. Thus, we can state that the energies of pairs of molecules A-A and pairs of molecules B-B are, respectively:

$$\varepsilon_{\mathrm{AA}} = -\frac{2\varepsilon_{\mathrm{A(A)}}}{z} \quad [2.43a]$$

$$\varepsilon_{\mathrm{BB}} = -\frac{2\varepsilon_{\mathrm{B(B)}}}{z} \quad [2.43b]$$

NOTE.– As we are essentially dealing with the mixing values, the pure solid model used has no influence on the results.

Let us define an energy, called the *exchange energy* w_{AB}, such that, if we start with two pure solids A and B, the exchange of a molecule of A and a molecule of B between the two solids increases the system's energy by $2w_{\mathrm{AB}}$. In the course of this process, we have destroyed z pairs A-A and z pairs B-B, and created $2z$ pairs A-B.

Thus, the average potential energy of the pairs A-B will be:

$$\varepsilon_{\mathrm{AB}} = \left(-2\varepsilon_{\mathrm{A(A)}} - 2\varepsilon_{\mathrm{B(B)}} + 2w_{\mathrm{AB}}\right)/2z = \left(-\varepsilon_{\mathrm{A(A)}} - \varepsilon_{\mathrm{B(B)}} + w_{\mathrm{AB}}\right)/z \quad [2.44]$$

In view of equations [2.43a], [2.43b] and [2.44], the exchange energy is, therefore:

$$w_{\mathrm{AB}} = z\left(\varepsilon_{\mathrm{AB}} - \frac{\varepsilon_{\mathrm{AA}} + \varepsilon_{\mathrm{BB}}}{2}\right) \quad [2.45]$$

The value w_{AB} will be independent of the temperature, provided the type of environment of a molecule, and therefore z, is not altered.

In addition, the different energies of interaction between atoms do not depend on the interatomic distances.

NOTE.– The energy of interaction between a molecule of A and a molecule of B in the solution can be written as:

$$\varepsilon_{\mathrm{AB}} = -2\varepsilon_{\mathrm{AB(AB)}} z = -2\varepsilon_{\mathrm{BB(AB)}}/z$$

so that, in light of relation [2.45], the exchange energy can be written as follows:

$$w_{\mathrm{AB}} = \varepsilon_{\mathrm{A(A)}} + \varepsilon_{\mathrm{B(B)}} - 2\varepsilon_{\mathrm{AB(AB)}} \quad [2.46]$$

We can establish a general form for the integral of configuration in the case of pairwise interactions.

A configuration is defined by the number of pairs A-A, pairs B-B and pairs A-B in the solid solution. For each configuration, there are several corresponding states, with each being defined by the distribution of the pairs in the spatial configuration.

The problem of modeling the solution boils down to being able to express the configurational partition function of the mixture, which involves accepting that the normal modes of vibration of the lattice are independent of the configuration. Thus, by definition, the configurational partition function is written as:

$$Z_c = \sum_i \exp\left(\frac{E_{I_i}}{\mathrm{k_B}T}\right) \qquad [2.47]$$

For each configuration, the solution is made up of a certain number of pairs A-A, pairs B-B and mixed pairs A-B, arranged in a very specific way in the space. E_{I_i} is the configurational energy of the state *i*. The above sum is extended to all configurations *i* for a given composition.

For a composition of N_A atoms of A and N_B atoms of B distributed randomly, the number of states is equal to the number of possible ways in which to distribute the N_A atoms of A and the N_B atoms of B on the $N(=N_\mathrm{A}+N_\mathrm{B})$ sites of the lattice – i.e.:

$$g = \frac{(N_\mathrm{A}+N_\mathrm{B})!}{N_\mathrm{A}!N_\mathrm{B}!} \qquad [2.48]$$

Let us consider a particular configuration of the solution in which the number of mixed pairs A-B is zX.

Thus, the number of neighbors of a molecule of A which are not molecules of B is $z(N_\mathrm{A}-X)$, so the number of pairs A-A is $z(N_\mathrm{A}-X)/2$, and by the same token, the number of pairs B-B is $z(N_\mathrm{B}-X)/2$.

The total number of pairs N_p of the three species is:

$$N_p = \frac{z}{2}\left(N_{\mathrm{AA}} + N_{\mathrm{BB}} + N_{\mathrm{AB}}\right) = \frac{z}{2}(N_{\mathrm{A}} + N_{\mathrm{B}}) \qquad [2.49]$$

For this particular configuration, characterized by the value of X, the potential energy of the solution, due to the interactions, will be:

$$E_{I_i} = \frac{z(N_{\mathrm{A}} - X)}{2}\left(-\frac{2\varepsilon_{\mathrm{A(A)}}}{z}\right) + \frac{z(N_{\mathrm{B}} - X)}{2}\left(-\frac{2\varepsilon_{\mathrm{B(B)}}}{z}\right) + zX\frac{(-\varepsilon_{\mathrm{A(A)}} - \varepsilon_{\mathrm{B(B)}} + \varepsilon_{\mathrm{AB}})}{z} \qquad [2.50]$$

Thus:

$$E_{I_i} = -N_{\mathrm{A}}\varepsilon_{\mathrm{A}} - N_{\mathrm{B}}\varepsilon_{\mathrm{B}} + Xw_{\mathrm{AB}} \qquad [2.51]$$

This relation can also be written differently, if we take account of expressions [2.43] and [2.48], as follows:

$$E_{I_i} = \frac{zN_{\mathrm{A}}\varepsilon_{\mathrm{AA}}}{2} + \frac{zN_{\mathrm{B}}\varepsilon_{\mathrm{BB}}}{2} + Xz\left(\varepsilon_{\mathrm{AB}} - \frac{\varepsilon_{\mathrm{AA}} + \varepsilon_{\mathrm{BB}}}{2}\right) \qquad [2.52]$$

The mixing term due to this configuration is, therefore, according to relation [2.45]:

$$E_c^{mix} = Xz\left(\varepsilon_{\mathrm{AB}} - \frac{\varepsilon_{\mathrm{AA}} + \varepsilon_{\mathrm{BB}}}{2}\right) = Xw_{\mathrm{AB}} \qquad [2.53]$$

We note in passing that, if x_{A} and x_{B} are, respectively, the molar fractions of the components A and B, the internal energy for a mole of solution ($N_{\mathrm{A}} + N_{\mathrm{B}} = \mathrm{N_a}$), which is also its enthalpy because the mixing volume is null, is:

$$U_m = H_m = \mathrm{N_a}\left[\frac{zx_{\mathrm{A}}\varepsilon_{\mathrm{AA}}}{2} + \frac{zx_{\mathrm{B}}\varepsilon_{\mathrm{BB}}}{2}\right] + Xz\left(\varepsilon_{\mathrm{AB}} - \frac{\varepsilon_{\mathrm{AA}} + \varepsilon_{\mathrm{BB}}}{2}\right) \qquad [2.54]$$

By substituting the expression of [2.53] back into equation [2.47], the mixing partition function becomes:

$$Z_c^{mix} = \sum_i g_i \exp\left(-\frac{Xw_{AB}}{k_B T}\right) \qquad [2.55]$$

In order to calculate the sum in expression [2.55], we define the value Y by the relation:

$$g \exp\left(-\frac{Yw_{AB}}{k_B T}\right) = \sum \exp\left(-\frac{Xw_{AB}}{k_B T}\right) \qquad [2.56]$$

This value is the fictitious number of mixed pairs which would yield the correct value of the partition function if the molecules were distributed at random. Thus, if we take account of relation [2.48], this mixing partition function Z_c^{mix} is:

$$Z_c^{mix} = \frac{(N_A + N_B)!}{N_A!N_B!} \exp\left(-\frac{Yw_{AB}}{k_B T}\right) \qquad [2.57]$$

The mixing partition function is the contribution of mixing to the overall partition function of the system.

We can now calculate the Helmholtz energy of mixing and the Gibbs energy of mixing, which are equal because the mixing volume is null, by:

$$\begin{aligned} G^{mix} = F^{mix} &= -k_B T \ln Z_c^{mix} = Yw_{AB} - k_B T \ln \frac{(N_A + N_B)!}{N_A!N_B!} \\ &= Yz\left(\varepsilon_{AB} - \frac{\varepsilon_{AA} + \varepsilon_{BB}}{2}\right) - k_B T \ln \frac{(N_A + N_B)!}{N_A!N_B!} \end{aligned} \qquad [2.58]$$

If we use Stirling's approximation, this Gibbs energy of mixing takes the form:

$$G^{mix} = F^{mix} = Yw_{AB}$$
$$= \mathrm{k_B}T\left[N_\mathrm{A}\ln\frac{N_\mathrm{A}}{(N_\mathrm{A}+N_\mathrm{B})} + N_\mathrm{B}\ln\frac{N_\mathrm{B}}{(N_\mathrm{A}+N_\mathrm{B})}\right] \quad [2.59]$$

Thus, we need to calculate the value of Y. In order to do so, we let $\langle X \rangle$ be the value of X at equilibrium at the given temperature. This value is that which gives a maximum of the partition function, and this value is identical to the average value which, according to the average theorem, is such that:

$$\sum X \exp\left(-\frac{w_\mathrm{AB}}{\mathrm{k_B}T}\right) = \langle X \rangle \sum \exp\left(-\frac{w_\mathrm{AB}}{\mathrm{k_B}T}\right) \quad [2.60]$$

By deriving equation [2.56] in relation to temperature, we can verify the expression:

$$\langle X \rangle = Y - T\left[\frac{\partial Y}{\partial T}\right] = \frac{\partial[Y/T]}{\partial(1/T)} \quad [2.61]$$

Thus, ultimately, the calculation of the partition function for our model of the solution consists of determining first $\langle X \rangle$ and then Y by using expression [2.61]. Finally, by substituting the value of Y thus obtained in expression [2.59], we obtain the value of the Gibbs energy of mixing.

2.3.2. *The microscopic model of the perfect solution*

Let us look again at the model given at the section 2.3.1, and determine the Gibbs energy of mixing by adding a sixth hypothesis to our model. Let us write that the exchange energy is null:

$$\varepsilon_\mathrm{AB} - \frac{\varepsilon_\mathrm{AA} + \varepsilon_\mathrm{BB}}{2} = 0 \quad [2.62]$$

Thus, by feeding this value back into relation [2.54], we see that the total enthalpy is equal to the sum of the enthalpies of the two pure solids, and therefore the enthalpy of mixing is null. This means that the mixing of our two solutions takes place without modification of the energy level – i.e. without a thermal effect.

With our new hypothesis, relation [2.57] becomes:

$$Z_{\mathrm{c}}^{mix} = \frac{(N_{\mathrm{A}} + N_{\mathrm{B}})!}{N_{\mathrm{A}}!N_{\mathrm{B}}!} \quad [2.63]$$

The Gibbs energy of mixing is identical to the enthalpy (zero mixing volume), and we calculate ($\mathrm{w}_{\mathrm{AB}} = 0$):

$$\begin{aligned} F^{mix} &= -\mathrm{k}_{\mathrm{B}} T \ln Z_{\mathrm{c}}^{mix} \cong \\ G^{mix} &\cong \mathrm{k}_B T \left(N_{\mathrm{A}} \ln \frac{N_{\mathrm{A}}}{N_{\mathrm{A}} + N_{\mathrm{B}}} + N_{\mathrm{B}} \ln \frac{N_{\mathrm{B}}}{N_{\mathrm{A}} + N_{\mathrm{B}}} \right) \end{aligned} \quad [2.64]$$

In order to find the molar value, we write $N_{\mathrm{A}} + N_{\mathrm{B}} = \mathrm{Na}$ and $n_{\mathrm{A}} + n_{\mathrm{B}} = 1$, so that $x_{\mathrm{A}} = n_{\mathrm{A}}$ and $x_{\mathrm{B}} = n_{\mathrm{B}}$. Thus, the molar Gibbs energy of mixing is:

$$G_m^{mix} \cong F_m^{mix} = \mathrm{R}T\left(x_{\mathrm{A}} \ln x_{\mathrm{A}} + x_{\mathrm{B}} \ln x_{\mathrm{B}}\right) \quad [2.65]$$

By derivation, we are able to calculate the partial Gibbs molar energies of mixing, and we note that the activity coefficients in reference (I) are equal to 1.

Hence, we have a model of the perfect solution, and note that this solution is not, as we might imagine, a solution in which the interactions between molecules are null, as is the case for a perfect gas, but rather is a solution where the energy of interaction $\varepsilon_{\mathrm{AB}}$ between two molecules of A and B is the arithmetic mean of the energies of the couples A-A and B-B because, from relation [2.62], we obtain:

$$\varepsilon_{AB} = \frac{\varepsilon_{AA} + \varepsilon_{BB}}{2} \qquad [2.66]$$

Of course, as is often the case with modeling, we have established one model of a perfect solution, but there is nothing to say that this is the only possible model.

2.3.3. *Microscopic model of strictly-regular solutions*

We will now go back to our model solution from section 2.3.1, with its six underlying hypotheses, but omit hypothesis 6, which is expressed by equation [2.62]. The exchange energy w_{AB} is neither null, nor it is sufficient to counteract the thermal agitation, therefore the random distribution is preserved.

NOTE.– As we continue to use the hypothesis of a random distribution of the molecules of A and B, this hypothesis appears to conflict with the existence of an exchange energy that means the minimum of the Helmholtz energy cannot correspond exactly to the random distribution. In other words, we circumvent the contradiction by accepting that the short-distance order, which would be introduced by the exchange energy, is annihilated by the thermal agitation. That is we accept the condition:

$$w_{AB} << k_B T \qquad [2.67]$$

Thus, let us write the random distribution of the mixed pairs A-B. Around a molecule of A, on average, there are zx_B molecules of B, so the average number of such pairs will be given by:

$$z\langle X\rangle = N_A z x_B = N_B z x_A \qquad [2.68]$$

This enables us to write:

$$\langle X\rangle^2 = N_A N_B x_A x_B = N_A N_B \left(1 - \frac{N_A}{N_A + N_B}\right)\left(1 - \frac{N_B}{N_A + N_B}\right) \qquad [2.69]$$

Hence:

$$\langle X\rangle^2 = \left(N_{AA} - \langle X\rangle\right)\left(N_{BB} - \langle X\rangle\right) \qquad [2.70]$$

Thus, $\langle X\rangle$ is independent of the temperature. This is known as the *zero-order approximation* or the *Bragg and William approximation.*

This temperature-independence of $\langle X\rangle$ means that, in view of relation [2.61], we can write:

$$\langle X\rangle = Y \qquad [2.71]$$

and, by using relation [2.68], we deduce:

$$\langle X\rangle = Y = \frac{N_A N_B}{N_A + N_B} \qquad [2.72]$$

The partition function of mixing is obtained on the basis of relation [2.57]. It will be the product of two terms: one identical to that in relation [2.63], relative to perfect solution, and the other constituted by the exponential, which will, in fact, be the excess term. According to equation [2.41], this term is:

$$G^{xs} = Yw_{AB} = \frac{\mathrm{N}_a n_A n_B}{n_A + n_B} w_{AB} \qquad [2.73]$$

and for the corresponding molar value ($n_A + n_B = 1$), we obtain:

$$G_m^{xs} = \mathrm{N}_a x_A x_B w_{AB} \qquad [2.74]$$

As Y is independent of the temperature, the excess entropy is null, and we find the following for the molar enthalpy:

$$H_m^{xs} = -T^2 \frac{\partial\left(G_m^{xs}/T\right)}{\partial T} = \mathrm{N}_a x_A x_B w_{AB} \qquad [2.75]$$

By deriving relation [2.73], we obtain the partial molar excess Gibbs energies, which give us the activity coefficients in reference (I), and we obtain:

$$\ln \gamma_{\mathrm{A}}^{(I)} = \frac{1}{\mathrm{R}T}\frac{\partial G^{xs}}{\partial n_A} = \frac{B}{T}x_{\mathrm{B}}^2 \quad [2.76]$$

$$-\mathrm{R}T \ln \gamma_{\mathrm{B}}^{(I)} = \frac{\partial G^{xs}}{\partial n_{\mathrm{B}}} = -Bx_{\mathrm{A}}^2 \quad [2.77]$$

with $B = \frac{\mathrm{N_a}\, w_{AB}}{\mathrm{R}}$ [2.78]

Thus, the difference between the strictly-regular solution and the perfect solution is simply the exchange energy, which is null for the latter.

2.3.4. *Microscopic model of the ideal dilute solution*

Let us look again at the above model of the strictly-regular solution (section 2.3.3). If the number of molecules in B is smaller than that of A – i.e. if the solution is very dilute in terms of B – then relation [2.72] gives us:

$$\langle X \rangle = Y \cong N_{\mathrm{B}} \quad [2.79]$$

Relation [2.73] yields:

$$G^{xs} = \mathrm{N_a} n_{\mathrm{B}} w_{\mathrm{AB}} \quad [2.80]$$

Thus, the activity coefficient of B in convention (I) satisfies the equation:

$$-\mathrm{R}T \ln \gamma_{\mathrm{B}}^{(I)} = \frac{\partial G^{xs}}{\partial n_B} = \mathrm{N_a} w_{\mathrm{AB}} \quad [2.81]$$

From this, we deduce:

$$\gamma_{\mathrm{B}}^{(I)} = \exp - \frac{\mathrm{N_a} w_{\mathrm{AB}}}{\mathrm{R}T} \qquad [2.82]$$

This activity coefficient is independent of the composition of the solution; it depends only on the temperature.

The excess Gibbs energy is, therefore:

$$G^{xs} = \mathrm{N_a} n_{\mathrm{B}} w_{\mathrm{AB}} = -\mathrm{R}T \ln \gamma_{\mathrm{B}}^{(I)} \qquad [2.83]$$

In that excess Gibbs energy, there is no contribution from the solvent. We can deduce from this that $\gamma_{\mathrm{A}}^{(I)} = 1$. Therefore, as is the case for the solvent, conventions (I) and (II) are identical, so we can deduce:

$$\gamma_{\mathrm{A}}^{(I)} = \gamma_{\mathrm{A}}^{(II)} = 1 \qquad [2.84]$$

In convention (II), the activity coefficient of the solute B is, therefore, also constant, and as is the case for the infinitely-dilute solution, the value of this constant is 1. Hence, we can deduce that:

$$\gamma_{\mathrm{B}}^{(II)} = 1 \qquad [2.85]$$

Thus, Henry's constant is given by:

$$K_{\mathrm{B}H} = \gamma_{\mathrm{B}}^{(I)} = \exp\left(-\frac{\mathrm{N_a} w_{\mathrm{AB}}}{\mathrm{R}T} \right) \qquad [2.86]$$

Thus, the ideal dilute solution is a specific case of the strictly-regular solution when one of the components is present only in very small proportions.

We deduce the physical meaning of the chemical potential in reference state (II) for the solute:

$$\mu_s^{\infty} = g_s^0 - \mathrm{N_a} w_{\mathrm{AB}} \qquad [2.87]$$

The chemical potential in reference state (II) is that of reference state (I) less the molar exchange energy.

We have just examined three models of solutions. The most commonly found is that of the strictly-regular solution. The other two models are two specific cases; that of the perfect solution if we make $w_{AB} = 0$, and that of the ideal dilute solution if we make $n_A >> n_B$. In all these models, we have only made the hypothesis of random distribution of the molecules, which is a fairly restrictive hypothesis, as mentioned with the strictly-regular solution. It is this hypothesis which we will examine in much detail in the next section.

2.3.5. *Fowler and Guggenheim's quasi-chemical model of the solution*

Let us look again at our binary solution formed of the two components A and B. The exchange energy is given by w_{AB}. This time, this energy is sufficient to influence the distribution of the atoms, which, therefore, is no longer random.

The quasi-chemical model of a solution postulates the existence of a true thermodynamic equilibrium[2] between one pair A-A, one pair B-B and two pairs A-B, as follows:

$$\text{A-A} + \text{B-B} = 2\text{A-B} \qquad [2R.1]$$

Such an equilibrium is known as a *quasi-chemical equilibrium.*

We will apply the law of mass action to that equilibrium, supposing that the species A-A, B-B and A-B constitute a perfect solution.

$$\frac{\left[\dfrac{z\langle X\rangle}{z\langle X\rangle+z\left(N_{\mathrm{A}}-\langle X\rangle\right)+z\left(N_{\mathrm{B}}-\langle X\rangle\right)}\right]^2}{\dfrac{z\left(N_{\mathrm{A}}-\langle X\rangle\right)}{z\langle X\rangle+z\left(N_{\mathrm{A}}-\langle X\rangle\right)+z\left(N_{\mathrm{B}}-\langle X\rangle\right)}\,\dfrac{z\left(N_{\mathrm{B}}-\langle X\rangle\right)}{z\langle X\rangle+z\left(N_{\mathrm{A}}-\langle X\rangle\right)+z\left(N_{\mathrm{B}}-\langle X\rangle\right)}}=K_{QC} \qquad [2.88]$$

2 A complete demonstration of this model, yielding this result, is given in Chapter 3 of Volume 2 of this series, on liquid phases (see Bibliography).

We calculate the value of the equilibrium constant using statistical thermodynamics. Because we assume that there is only one mode of vibration of the lattice, which is identical for each of the three solids (pure A, pure B and the solid solution), the equilibrium constant is reduced to the exponential of the difference in energy between the initial state and the final state. This difference is -2w$_{AB}$. Thus, we obtain:

$$\langle X\rangle^2 = \left(N_A - \langle X\rangle\right)\left(N_B - \langle X\rangle\right)\exp - \frac{2w_{AB}}{k_B T} \quad [2.89]$$

A laborious calculation, as done by Christian in 1975 [CHR 75], gives us:

$$\langle X\rangle = \frac{2N_A N_B}{\left(N_A + N_B\right)\beta_q + 1} \quad [2.90]$$

If we introduce the molar fractions defined by:

$$x_A = \frac{n_A}{\left(n_A + n_B\right)} = \frac{N_A}{\left(N_A + N_B\right)} \quad [2.91a]$$

and $x_B = \dfrac{n_B}{\left(n_A + n_B\right)} = \dfrac{N_B}{\left(N_A + N_B\right)}$ [2.91b]

for a mole of mixture $\left(N_A + N_B = \mathrm{N_a}\right)$, we have:

$$\langle X\rangle = \frac{2\mathrm{N}_a^2 x_A x_B}{\mathrm{N}_a \beta_q + 1} \quad [2.92]$$

The term "β_q" is defined by:

$$\beta_q = \frac{\left\{1 + 4N_A N_B\left[\exp - \frac{2w_{AB}}{z k_B T} - 1\right]\right\}^{1/2}}{N_A + N_B} \quad [2.93]$$

This can also be written as:

$$\beta_q = \left\{ \left(1-2x_A\right)^2 + 4x_A x_B \exp - \frac{2w_{AB}}{zk_B T} \right\}^{1/2} \qquad [2.94]$$

By substituting expression [2.92] into equation [2.61], we find:

$$Y = \frac{N_A N_B}{N_A + N_B} \frac{zk_B T}{2w_{12}} \int_0^{2w_{AB}/zk_B T} \frac{2}{\beta_q - 1} \mathrm{d}\left(\frac{2w_{AB}}{zk_B T}\right) \qquad [2.95]$$

The lower bound of the integral of expression [2.95] is determined as follows: if $T \to \infty$, we must find a random distribution (because then $w_{12} << k_B T$) in which Y is basically equal to $\langle X \rangle$, according to relation [2.59], and the difference $Y - \langle X \rangle$ is finite. Thus, Y / T tends toward zero as $1/T$ tends toward zero. In view of the definition of β_q (relation [2.94]), we have:

$$\exp\left(\frac{2w_{AB}}{zk_B T}\right) = \frac{\beta_q^2 - (1-2x_A)}{4x_A x_2} \qquad [2.96]$$

Thus:

$$\mathrm{d}\left(\frac{2w_{AB}}{zk_B T}\right) = \frac{2\beta_q \,\mathrm{d}\beta_q}{\left(\beta_q - 1 + 2x_A\right)\left(\beta_q + 1 - 2x_A\right)} \qquad [2.97]$$

By introducing this in expression [2.95], we find:

$$\frac{Y}{N_A N_B} = \frac{zk_B T}{2w_{AB}} \left(x_A \ln \frac{\beta_q - 1 + 2x_A}{2x_A} + x_B \ln \frac{\beta_q + 1 - 2x_A}{2x_B} - \ln \frac{\beta_q + 1}{2} \right) \qquad [2.98]$$

in the knowledge that:

$$(N_A + N_B = 1) \qquad [2.99]$$

By substituting this back into relation [2.58], we obtain the excess molar Gibbs energy:

$$G^{xs} = \frac{1}{2} zRT \left[x_{B} \ln \frac{\beta_q + x_B - x_A}{x_B(\beta_q + 1)} + x_A \ln \frac{\beta_q + x_A - x_B}{x_A(\beta_q + 1)} \right] \qquad [2.100]$$

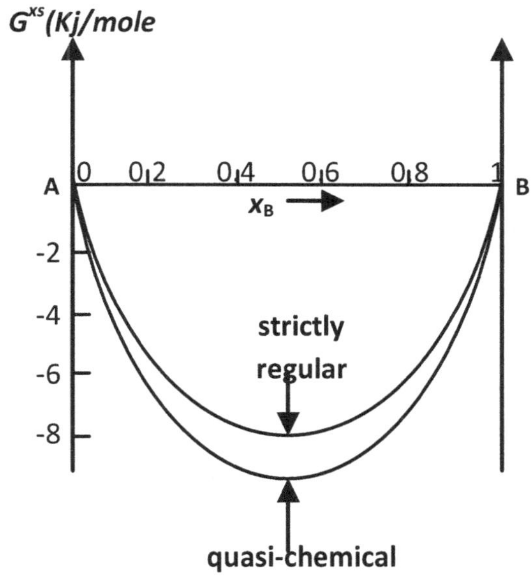

Figure 2.11. *Comparison of the excess Gibbs energy values for a strictly-regular solution and for the quasi-chemical model (data from [DES 10])*

We note that, as it is the case for strickly-regular solutions, the model is symmetric relative to the two components A and B.

Figure 2.11 shows the gap between the excess Gibbs molar energy of the strictly-regular solution and that of this model, for the values T = 800 K, z = 12 and $N_a w_{AB}$ = 30 kJ. The two curves exhibit a minimum at $x_A = x_B = 0.5$.

As we can see, the solution is no longer regular.

The activity coefficients of the two components become, by derivation of the excess Gibbs energy in relation to n_i, remembering that β_q, x_A and x_B are the functions of N_A and N_B (the calculation is complex but is direct):

$$\gamma_{A}^{(I)} = \left(\frac{\beta_q + x_A - x_B}{x_A \left(\beta_q + 1 \right)} \right)^{z/2} \quad [2.101]$$

$$\gamma_{B}^{(I)} = \left(\frac{\beta_q + x_B - x_A}{x_B \left(\beta_q + 1 \right)} \right)^{z/2} \quad [2.102]$$

The solution thus modeled is called the *quasi-chemical model.*

This model can be calibrated on an experiment with a single adjustable parameter β_q. If we know the values of the activity coefficients at a given temperature T, for a given composition, using relations [2.101] and [2.102], we calculate $G_{\grave{u}}^{xs}$; then, with relation [2.98] we deduce β_q, and with equation [2.94] we obtain the ratio w_{AB}/z, which is the difference $\varepsilon_{AB} - \frac{\varepsilon_{AA} + \varepsilon_{BB}}{2}$.

NOTE.– We have seen that relation [2.70] of the strictly-regular solution model was called the Bragg and Williams zero-order approximation. Similarly, in view of relation [2.89], the quasi-chemical solution is called an approximation of order 1. The order at hand is, in fact, the power to which the exponential appearing in relation [2.89] is raised: power zero for the Bragg and Williams model, and power 1 for the quasi-chemical model.

Let us calculate Warren and Cowley's parameter for the quasi-chemical model.

If we try to link our results to the properties introduced in section 2.2.1, then we immediately find:

$$z'_A = z'_B = z \tag{2.103}$$

$$\langle X \rangle = N_A y_{AB} = N_A y_{BA} = N_B x_A = N_A x_B = N_{AB} / z \tag{2.104}$$

$$\eta_{AB} = \eta_{BA} = \eta = 1 - \frac{x_{AB}}{x_A x_B} \tag{2.105}$$

By substituting this relation back into equations [2.90] and [2.96], we obtain:

$$\eta = 1 - \frac{2zx_A x_B}{\beta_q - 1} \tag{2.106}$$

β_q is, as ever, defined by relation [2.94].

Thus, the activity coefficients expressed by relations [2.101] and [2.102] can be written in the form:

$$\ln \gamma_A = \frac{z}{2} \ln\left(\frac{1 - x_B(1-\eta)}{x_A}\right) \tag{2.107a}$$

$$\ln \gamma_B = \frac{z}{2} \ln\left(\frac{1 - x_A(1-\eta)}{x_B}\right) \tag{2.107b}$$

Figure 2.12 shows the variations in Warren and Cowley's order parameter for the quasi-chemical model, with the following values of the parameters: T = 800 K, $N_a w_{AB}$ = 30 kJ and z = 12. The result obtained respects the symmetry of the model, with a minimum for the composition $x_A = x_B = 0.5$.

2.4. Thermodynamic study of the degree of order of an alloy

The concept of a long-distance degree of order leads us to ask the question: what would the degree of order of a solid solution be in given conditions of temperature and composition?

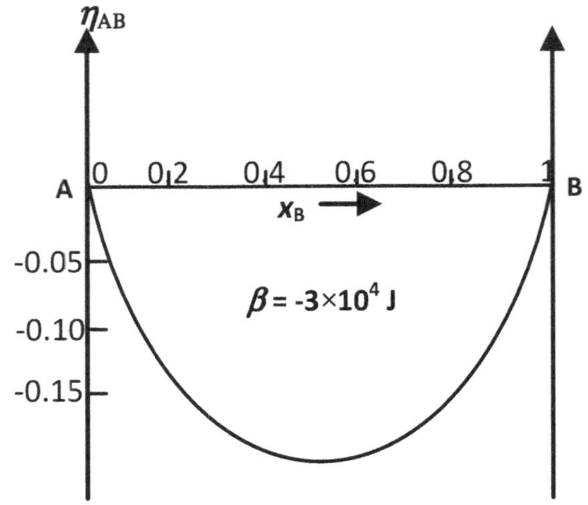

Figure 2.12. *Variation of the degree of order as a function of the composition of a binary solution in the quasi-chemical model (data taken from [DES 10])*

The method used to calculate this degree of order consists of expressing the term of the Helmholtz energy due to the configuration of the mixture of atoms as a function of the degree of order. The Helmholtz energy is calculated on the basis of the configuration term in the partition function, which is also a function of the degree of order. The state of equilibrium – i.e. the value of the degree of order for a certain temperature – is then found by making the derivative of the configurational Helmholtz energy in relation to the degree of order equal to zero.

2.4.1. *Hypotheses of the model: configuration energy*

We suppose that there is no short-distance order in the alloy, and therefore all of the configurational partition function is due to the long-distance order.

All of the configuration energy is due solely to an interaction between near-neighboring atoms, and each atom A or B is surrounded by z near neighbors. The energies of the pairs AA, BB and AB are different.

For a given composition of the solution and in a given state τ, we would have n_{AA} pairs A-A with energy, ε_{AA}, n_{BB} pairs B-B with energy ε_{BB} and n_{AB} pairs A-B whose energy is ε_{AB}. For this state τ, the configuration energy will, therefore be:

$$E_\tau = \left(n_{AA}\varepsilon_{AA} + n_B\varepsilon_{BB} + n_{AB}\varepsilon_{AB}\right) \qquad [2.108]$$

Using relation [2.45], this expression takes the form:

$$E_\tau = \left[\left(n_{AA} + \frac{n_{AB}}{2}\right)\varepsilon_{AA} + \left(n_{BB} + \frac{n_{AB}}{2}\right)\varepsilon_{BB} + n_{AB}\frac{w_{AB}}{z}\right] \qquad [2.109]$$

In the models developed below, the interaction energies will be taken to be independent of both the temperature and the interatomic distances or, if readers prefer, we suppose that these interatomic distances are not altered by the creation of order.

2.4.2. *Expression of the configuration partition function*

The configuration partition function is a function of the temperature which can be expressed as a sum of terms, each of which is the contribution that corresponds to a value of the degree of order s. We express this idea as follows:

$$Z_C(T) = \sum_s Z_C(T,s) \qquad [2.110]$$

Let us explicitly write each of these contributions, which are characterized by a value E_τ of the energy expressed by equation [2.108], so by definition of the partition function:

$$Z_C(T,s) = \sum_\tau \exp - \frac{E_\tau}{k_B T} \qquad [2.111]$$

For a given degree of order and a given temperature, as all the configurations have the same energy, the terms in the above sum are all identical, and if we let $g(s)$ denote the number of those configurations, the partition function of configuration will be:

$$Z_C(T,s) = g(s)\exp - \frac{E(s)}{\mathrm{k_B}T} \quad [2.112]$$

In order to evaluate the functions $g(s)$ and $E(s)$, we need to know the distribution of the atoms on the lattice for the given value of s. Two models have been developed: the Gorsky, Bragg and Williams model and the quasi-chemical model. The hypotheses upon which these models are based are similar, respectively, to those used for the model of a strictly-regular solution (see section 2.3.3) and those used for Fowler and Guggenheim's quasi-chemical solution model (see section 2.3.5).

We are going to look at these models in the simple case of an alloy where the molar fractions of A and B are equal to ½ and the fractions of sites are also equal to ½ (that is $x = y = ½$ in the notation used in Table 2.3) – this is the Gorsky expansion. The calculations are carried out in the same way for any given values of x and y, but the expansions are, naturally, more complex.

2.4.3. *The Gorsky, Bragg and Williams model*

The Gorsky, Bragg and Williams model is denoted by the abbreviation GBW. This model postulates that between the atoms of the alloy, there are pairwise interactions over short distances, which are only experienced by the nearest neighbors. We have already used these interactions in section 2.3.1 and the energies in expressions [2.43–2.45]. However, as in the case of strictly-regular solutions, we suppose that these interactions have no influence on the relative arrangement of the atoms, which remains random. For a value of s, we have seen in the last row of Table 2.3 that the number of atoms of A on a site A is $N(1+s)/4$. Each of them has z sites for B as near neighbors, and the average fraction of occupation of these z sites by an atom of B is

(1-s)/2. The number of pairs A-A in the solution will, therefore, be given by:

$$n_{\text{AA}} = \frac{Nz}{8}\left(1-s^2\right) \qquad [2.113]$$

In the same way, we calculate the number of pairs B-B, which gives us the same result. The number of A-B pairs will thus be given by the difference between the total number of pairs (which is $Nz/2$) and the number of pairs A-A and B-B:

$$n_{\text{AB}} = \frac{Nz}{2} - n_{\text{AA}} - n_{\text{BB}} \qquad [2.114]$$

Table 2.4 recalls the number of pairs of the different types. On the basis of the data given in Table 2.4 and relation [2.109], we can easily calculate the configuration energy of mixing, which is, therefore, an average energy, for each value of s. We obtain:

$$< E(s) >= \frac{Nz}{8}\left(\varepsilon_{\text{AA}} + \varepsilon_{\text{BB}} + 2\varepsilon_{\text{AB}} + \frac{2w_{\text{AB}}s^2}{z}\right) \qquad [2.115]$$

Type of pair	A-A	B-B	A-B
Number	$\frac{Nz}{8}\left(1-s^2\right)$	$\frac{Nz}{8}\left(1-s^2\right)$	$\frac{Nz}{4}\left(1+s^2\right)$

Table 2.4. *Number of pairs of different types for a random distribution in an alloy AB*

If we let $< E(0) >$ be the average energy if s = 0, then relation [2.115] becomes:

$$< E(s) >=< E(0) > + \frac{Nw_{\text{AB}}s^2}{4} \qquad [2.116]$$

Let us now examine how to calculate the statistical weight. We said earlier that the atoms of A and B were distributed at random, and therefore the statistical weight will be the number $g_a(s)$ equal to the number of ways there are in which to randomly divide the $N/2$ sites A into two groups and the $N/2$ sites B into two groups. The number of sites of each group is given by the last row in Table 2.3 ($x = y = 1/2$). Hence, we can write:

$$g_a(s) = \left[\frac{N_{(A)}!}{N_{A_A}!N_{B_A}!}\right]^2 = \frac{(N/2)!(N/2)!}{\left\{\left[\frac{N}{4}(1+s)\right]!\right\}^2\left\{\left[\frac{N}{4}(1-s)\right]!\right\}^2} \quad [2.117]$$

By switching to logarithms and using Stirling's approximation, we obtain:

$$\ln g_a(s) = -\frac{N}{2}\left[(1+s)\ln(1+s)+(1-s)\ln(1-s)\right]-2\ln 2 \quad [2.118]$$

We can now calculate the Helmholtz energy of configuration based on relation [2.112], using the expression:

$$F(s) = -k_B T \ln Z_C(T,s) = -k_B T \ln g_a(s) + < E(s) > \quad [2.119]$$

Into that expression of the Helmholtz energy, we feed the value of the statistical weight, the logarithm for which is given by relation [2.118], and the value of the energy given by relation [2.116]. We obtain:

$$F(s) = \frac{N}{2}k_B T\left[(1+s)\ln(1+s)+(1-s)\ln(1-s)\right]-2\ln 2 + < E(0) > + \frac{N w_{AB} s^2}{4} \quad [2.120]$$

We calculate the particular value obtained by that partition function for $s = 0$, thus:

$$F(0) = \frac{N}{2}k_B T\left[-2\ln 2\right] + < E(0) > \quad [2.121]$$

In order to eliminate the term <E(0)>, we work on the difference $2\frac{F(s)-F(0)}{N\text{k}_{\text{B}}T}$, which is written as:

$$2\frac{F(s)-F(0)}{N\text{k}_{\text{B}}T}=\left[(1+s)\ln(1+s)+(1-s)\ln(1-s)\right]-s^2 \quad [2.122]$$

Now, we look for the value of the degree of order at equilibrium of the system at temperature T. This value is obtained by zeroing the derivative of the Helmholtz energy in relation to s, or, which amounts to the same thing, by zeroing the derivative of the function [2.122]. We are led to:

$$\ln\frac{(1+s)}{(1-s)}=-\frac{w_{\text{AB}}s}{\text{k}_{\text{B}}T} \quad [2.123]$$

Let us recall the mathematical relation deduced from the definition of the hyperbolic tangent:

$$\ln\frac{(1+x)}{(1-x)}=\text{Arg}\tanh x \quad [2.124]$$

Thus, we find the solution at thermodynamic equilibrium:

$$s=\tanh\left(-\frac{w_{\text{AB}}}{2\text{k}_{\text{B}}T}\right) \quad [2.125]$$

For all values of the ratio $-w_{\text{AB}}/\text{k}_{\text{B}}T$, one of the roots of equation [2.125] is zero:

– when $-w_{\text{AB}}$ is negative, the root $s = 0$ is the only one which corresponds to a minimum of $F(s)$. Therefore, there is never long-distance order in the solution, and the alloy simply behaves like a strictly-regular solution as defined in section 2.3.3;

– when $-w_{AB}$ is positive, for high temperatures (i.e. for low values of $-w_{AB}/k_BT$), the root $s = 0$ is the only one which gives a minimum value for the function $F(s)$. There is still no long-distance order and the solution is, still, a simple strictly-regular solution;

– when $-w_{AB}$ is positive, for sufficiently low temperatures (i.e. for high values of $-w_{AB}/k_BT$), then in addition to the root $s = 0$, there is a second root whose value is between 0 and 1. By looking at the sign of the second derivative of the function $F(s)$, we see that the first root $s = 0$ corresponds to a maximum. Thus, the system is in an unstable state of equilibrium. The second root corresponds to a minimum of the function $F(s)$ and therefore a stable thermodynamic equilibrium.

In order to represent these variations in graph form, in Figure 2.13 we have shown the variations of the function $2\frac{F(s)-F(0)}{Nk_BT} = f(s)$ for different values of the ratio $-w_{AB}/k_BT$.

From this figure, it is clear that there is a critical value of the temperature Tc which is such that:

If $T < Tc$, the degree of order is positive, and decreases as the temperature rises. This degree reaches zero at the critical temperature and remains at zero when the temperature is higher.

For the critical temperature, both roots of the function [2.125] are equal to zero.

Hence, the critical temperature is determined by the threefold condition:

$$s = 0 \quad [2.126a]$$

$$\frac{\partial F}{\partial s} = 0 \quad [2.126b]$$

$$\frac{\partial^2 F}{\partial s^2} = 0 \quad [2.126c]$$

This gives us the expression of the critical temperature:

$$k_B T_c = -\frac{w_{AB}}{2} \qquad [2.127]$$

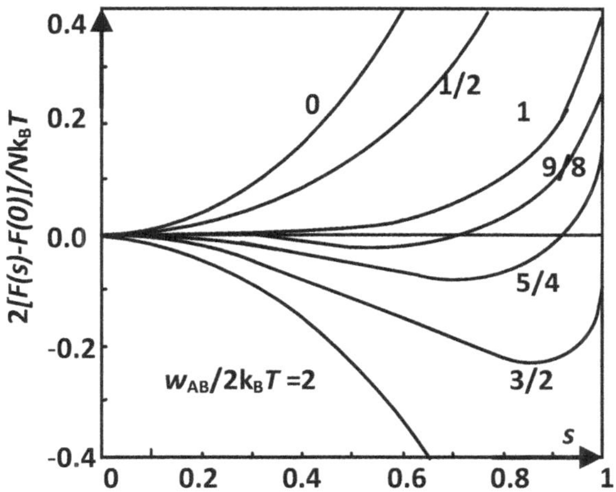

Figure 2.13. *Variations in Helmholtz energy as a function of the degree of order, according to the GBW model for a solid AB with different values of the ratio* $-w_{AB}/2k_BT$

This enables us to write the definition of s on the basis of relation [2.125] in the form:

$$\frac{s}{\operatorname{Arg}\tanh s} = \frac{T}{T_c} \qquad [2.128]$$

Figure 2.14(a) shows the variations in the degree of order for temperatures below the critical temperature.

Throughout our presentation of the GBW model, we have supposed that the composition of the alloy was $x = y = ½$. The same calculations can be performed, maintaining the equality of $x = y$, but with arbitrary atomic fractions $x_A = x$ and $x_B = 1\text{-}x$.

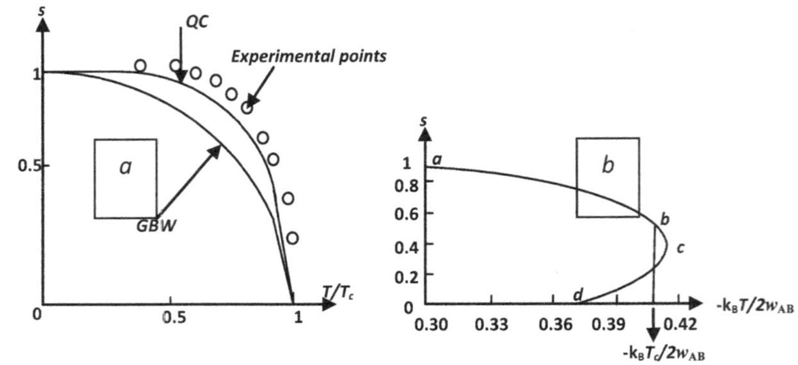

Figure 2.14. *Variations of the degree of order with temperature: a) solid of type AB and b) solid of type A_3B*

In Figure 2.14(b), we have shown the variations in the degree of order of system as a function of the quantity $-k_BT/2w_{AB}$, in the case of an A_3B solid – i.e. one where $x_A = 0.75$ and $x_B = 0.25$. As we can see, when the temperature is lowered, a second minimum of the function $F(s)$ appears for a value $s = 0$ of the degree of order. The first minimum is still at $s = 0$. The position of this second minimum descends as the temperature decreases. The first minimum corresponds to a metastable state, represented by the branch *ba* in Figure 2.14(b), and the second minimum corresponds to the state of stable thermodynamic equilibrium, represented by the branch *ab*. Between the two minima, obviously, there is a maximum which corresponds to an unstable state. This state corresponds to the branch *bc* in Figure 2.14(b).

Ultimately, in this case, at high temperatures the order parameter is zero. When the temperature decreases, we observe the emergence of a metastable ordered state (the branch *bc* on the curve), in addition to the stable ordered state (the branch *ab*).

2.4.4. *The quasi-chemical model*

The quasi-chemical model is denoted by the symbol "QC". We will now use a method similar to that which we used in section 2.3.1.

We will conduct the calculation in the simple case of the alloy AB, in which the atomic fractions and the fractions of sites are both equal to ½. More complex calculations, which are nevertheless similar on all points to those carried out above, can be used to solve more general cases.

In the circumstances of our scenario, we again refer to the last row of Table 2.3 for the expressions of the numbers of types of atoms as a function of the degree of order.

On the basis of Table 2.5, we will now calculate the number of pairs of neighboring sites occupied in the four possible cases: A_A-B_B, A_A-A_B, A_A-B_A and B_A-B_B for any value of the degree of order s. The total number of pairs, obviously, is $Nz/2$. If we let zX represent the number of pairs A_A-A_B, it is easy to fill in the cells of Table 2.5.

Type of pair	A_A-A_B	A_A-B_B	B_A-A_B	B_A-B_B
Number	zX	$z\left(N_{A_A}-X\right)$	$z\left(N/2-N_{A_A}-X\right)$	zX

Table 2.5. *Number of pairs of different types for an alloy AB*

It is clear that to set the value N_{A_A} is to set the value of s.

We can also see that if the two values N_{A_A} and X are known, then the number of pairs of atoms of each type in the solid is perfectly well determined. Therefore, we can bring together all the states which have the same values of those two properties N_{A_A} and X.

According to relation [2.109], the energy of such a state is:

$$E(X,N_{A_A})=z\frac{N}{2}\varepsilon_{AB}-2Xw_{AB}=E(0)-2Xw_{AB} \qquad [2.129]$$

In view of relation [2.112], we know that the configurational partition function for all the states characterized by a pair of values (X,N_{A_A}) is:

$$Z_C(T) = \sum_{\tau} g(X, N_{A_A}) \exp - \frac{E(X, N_{A_A})}{k_B T}$$ [2.130]

The problem of calculating this partition function is that we are unable to calculate the function $g(X, N_{A_A})$, which is the number of possible ways to distribute $Nz/2$ entities (pairs of atoms) into four groups containing the number of pairs indicated in Table 2.5. Indeed, we cannot consider that these pairs are distributed at random because the pairs of atoms are not independent entities. In order to show this, let us choose four sites – *a*, *b*, *c* and *d* – in the same plane of the crystalline lattice (see Figure 2.15). We place one pair of atoms A and B on sites *a* and *b*, one pair A-A on the sites *ac* and one pair B-B on the sites *bd*. It is then clear that the pair in place on the sites *cd* can only be a pair A-B.

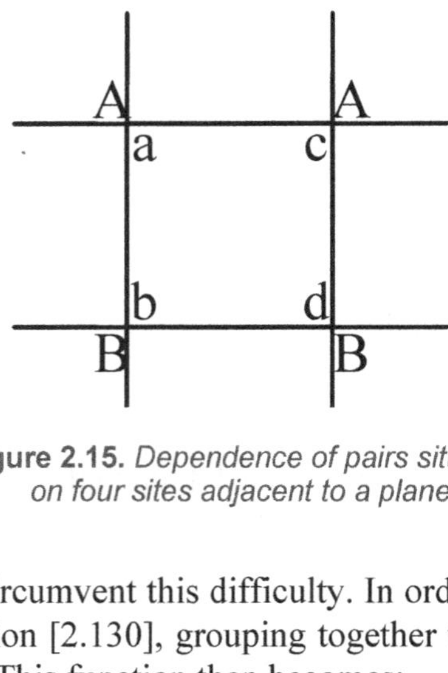

Figure 2.15. *Dependence of pairs situated on four sites adjacent to a plane*

We will now circumvent this difficulty. In order to do so, we write the partition function [2.130], grouping together those terms whose X value is the same. This function then becomes:

$$Z_C(T) = \sum_{N_{A_A}} \sum_{X} g(X, N_{A_A}) \exp - \frac{E(X, N_{A_A})}{k_B T}$$ [2.131]

The first sum is extended to all the values of X for a given value of N_{A_A}. We replace each by its maximum term. If we adopt N_{A_A}, for example, the maximum of the sum in relation to X would be determined if X satisfied the condition:

$$\frac{\partial \ln\left[g(X,N_{A_A})-E(X,N_{A_A})\right]/\mathrm{k_B}T}{\partial X}=0 \qquad [2.132]$$

We will assume that this condition is satisfied if X takes the value $\langle X\rangle$ and that this value results from a quasi-chemical equilibrium, which would be written as:

$$A_A\text{-}A_B + B_A\text{-}B_B = A_A\text{-}B_B + B_A\text{-}A_B \qquad [2R.2]$$

The pairs solution is then considered to be a perfect solution – i.e. one which satisfies the statistical weight $g_a(s)$ given by relation [2.117], and corresponds to a random distribution of the atoms of A and B.

This equilibrium expresses the exchange of two atoms between two sites and therefore involves the energy $2w_{AB}/z$.

The application of the law of mass action to the equilibrium [2R.2] gives us:

$$\frac{[B_A-A_B][A_A-B_B]}{[A_A-A_B][B_A-B_B]}=K \qquad [2.133]$$

We are dealing with the solid phase, so the only contributions to the partition functions of the components are the vibrational terms. If we suppose that we have an Einsteinian solid with a single fundamental frequency of vibration, independent of the distribution of the atoms, then the constant K calculated by statistical thermodynamics becomes:

$$K=\exp-\frac{2w_{AB}}{z\mathrm{k_B}T} \qquad [2.134]$$

By replacing the concentrations of the pairs with the values given in Table 2.5 (the volume being considered to be independent of the composition), we find:

$$\left[N_{\mathrm{A_A}} - <X>\right]\left[\frac{N}{2} - N_{\mathrm{A_A}} - <X>\right] = (<X>)\exp-\frac{2w_{\mathrm{AB}}}{z\mathrm{k_B}T} \qquad [2.135]$$

Using the data in Table 2.3, we deduce from this:

$$<X> = \frac{N}{2}\frac{\left(1-s^2\right)/2}{\sqrt{1+\left(1-s^2\right)\left(\exp-\frac{2w_{\mathrm{AB}}}{z\mathrm{k_B}T}-1\right)}+1} \qquad [2.136]$$

To simplify the formulation, we will set:

$$\beta_s = \sqrt{1+\left(1-s^2\right)\left(\exp-\frac{2w_{\mathrm{AB}}}{z\mathrm{k_B}T}-1\right)} \qquad [2.137]$$

The energy $<E(s)>$ corresponding to the average distribution <X> will then be as follows, in light of relation [2.129]:

$$<E(s)> = \frac{Nz}{2}\varepsilon_{\mathrm{AB}} - \frac{Nw_{\mathrm{AB}}}{2}\frac{\left(1-s^2\right)}{\beta_s+1} \qquad [2.138]$$

However, this function $<E(s)>$ corresponds to the average value of the energy as given by the mean value theorem, meaning that we have:

$$\sum_{\tau} E_{\tau}\exp-\frac{E_{\tau}}{\mathrm{k_B}T} \cong <E(s)> \sum_{\tau}\exp-\frac{E\tau}{\mathrm{k_B}T} \qquad [2.139]$$

We will now introduce a fictitious value of the energy, written as $<<E(s)>>$, so that with this value, the solution exhibits perfect behavior. This fictitious value, therefore, must satisfy the equation:

$$Z_C(\tau,s)=\sum_{\tau}\exp-\frac{E_\tau}{\mathrm{k_B}T}\cong g_a(s)\exp-\frac{<<E(s)>>}{\mathrm{k_B}T} \qquad [2.140]$$

By comparing expressions [2.132] and [2.140], we discover a relation between the two values $<E(s)>$ and $<<E(s)>>$, which is written as:

$$<E(s)>=\frac{\partial\left(<<E(s)>>/T\right)}{\partial\left(1/T\right)} \qquad [2.141]$$

Hence, by using this relation, the value of $<<E(s)>>$ which corresponds to the value of $<E(s)>$ given by expression [2.138] will be:

$$<<E(s)>>=\frac{N}{2}\left[z\varepsilon_{\mathrm{AB}}-\frac{w_{\mathrm{AB}}}{2}+\frac{z\mathrm{k_B}T}{2}\int_0^{\frac{-2w_{\mathrm{AB}}}{z\mathrm{k_B}T}}\frac{1-s^2}{\beta_s-1}d\left(\frac{-2w_{\mathrm{AB}}}{z\mathrm{k_B}T}\right)\right] \qquad [2.142]$$

The integration constant is determined by the condition:

$$\lim_{w_{\mathrm{AB}}\to 0}<<E(s)>>=<E(0)> \qquad [2.143]$$

The calculation of the above integral thus gives us:

$$<<E(s)>>=\frac{N}{2}\left\{\begin{array}{l}z\varepsilon_{\mathrm{AB}}-\dfrac{w_{\mathrm{AB}}}{2}\\ +\dfrac{z}{2}\left[(1+s)\ln\dfrac{\beta_s+s}{1+s}+(1-s)\ln\dfrac{\beta_s-s}{1-s}-2\ln\dfrac{\beta_s+1}{2}\right]\end{array}\right\} \qquad [2.144]$$

The configurational Helmholtz energy then becomes:

$$F=-\mathrm{k_B}T\ln g_a(s)+<<E(s)>> \qquad [2.145]$$

Hence, in view of relations [2.144] and [2.118]:

$$\begin{aligned}\frac{F(s)}{\frac{N}{2}\mathrm{k_B}T} &= (1+s)\ln(1+s)+(1-s)\ln(1-s)\\ &-2\ln 2+\frac{z}{2}\left\{\left[(1+s)\ln\frac{\beta_s+s}{1+s}+(1-s)\ln\frac{\beta_s-s}{1-s}-2\ln\frac{\beta_s+1}{2}\right]\right\}\end{aligned} \qquad [2.146]$$

We can easily calculate $F(0)$ and therefore the ratio:

$$\begin{aligned}2\frac{F(s)-F(0)}{N\mathrm{k_B}T} &= (1+s)\ln(1+s)+(1-s)\ln(1-s)\\ &+\frac{z}{2}\left\{\left[(1+s)\ln\frac{\beta_s+s}{1+s}+(1-s)\ln\frac{\beta_s-s}{1-s}-\ln\beta_s\right]\right\}\end{aligned} \qquad [2.147]$$

The function [2.147] generates graphic representations that are very similar to those shown in Figure 2.13.

By deriving the function [2.147] in relation to s and zeroing the derivative, we obtain the value of s at equilibrium, which is given by the implicit equation:

$$\ln\frac{\beta_s+s}{\beta_s-s}=\frac{z-2}{2}\ln\frac{(1+s)}{(1-s)} \qquad [2.148]$$

The critical temperature is then determined by applying conditions [2.126a]–[2.126c]. These conditions are fulfilled if $s = 0$ and:

$$\frac{w_{\mathrm{AB}}}{z\mathrm{k_B}T_c}=-z\ln\frac{z}{z-2} \qquad [2.149]$$

With Figure 2.14(a), we can compare the curve given by the GBW approximation with that given by the quasi-chemical approximation and the experimental result in the case of the alloy CuZn. We can see that the second curve is closer to the experimental points.

The quasi-chemical model is based on the hypothesis of distribution of the pairs represented by the equilibrium [2R.2]. This hypothesis has been proven by Bethe, using the grand partition function.

2.4.5. *Comparison of the models against experimental results*

The difficulty of the experimental approach is in actually achieving equilibrium. We need to make sure that the degree of order obtained is indeed that which corresponds to equilibrium at the chosen temperature. In order to do so, we usually proceed as follows: we heat the alloy to a high temperature, far greater than the critical temperature, so as to rapidly create total disorder in the alloy. The sample is then plunged as quickly as possible to a temperature lower than that at which we want to achieve the equilibrium. In most of the cases, although there are exceptions, at low temperature the alloy keeps the maximum disorder which it had previously acquired. Thus, we take a reading at the desired temperature and wait a sufficient amount of time to ensure that the equilibrium is established.

The degree of order is measured by X-ray diffraction. Indeed, the move from order to disorder is usually accompanied by a rise in the level of symmetry in the crystal, and the two arrangements of the atoms of A and B create substructures characterized by specific diffraction lines. However, as X-rays only give us an average picture of a situation, each experiment leaves room for a certain amount of doubts. Does the result represent the presence of ordered zones in a disordered matrix? Or does it reveal a value of the uniform degree of order in the sample? The answer is all the more delicate when the degree of order is lower. Of course, X-ray pictures can be taken at the temperature of the reading at equilibrium. Their stability over time constitutes an indicator that equilibrium has been reached.

The models and experiments can be compared on the basis of several values: the curve giving the degree of order as a function of the temperature, the measurement of the enthalpy of transformation, that of the critical temperature and its variations with the composition of the alloy, and finally the comparison of the calculated curves and the experimental curves illustrating the specific heat capacity as a function of the temperature.

2.4.5.1. *Variation of the degree of order with temperature*

In Figure 2.14(a), we saw the comparison of the curves showing the variations in the degree of order with temperature in the case of a solid AB – specifically, the alloy CuZn. We can see that the results of the two models are not too far from reality, but the QC model gives a better representation of the real world than does the GBW model.

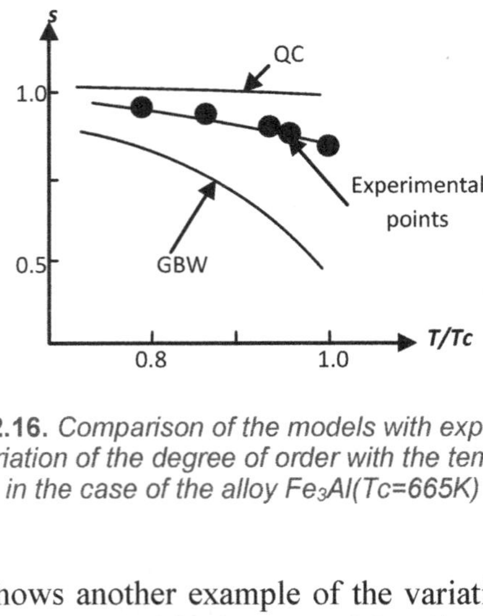

Figure 2.16. *Comparison of the models with experiments in the variation of the degree of order with the temperature in the case of the alloy Fe_3Al(Tc=665K)*

Figure 2.16 shows another example of the variation of the degree of order with the temperature in the case of the alloy Fe_3Al. We know that, in spite of its chemical formula, this alloy is an AB alloy, because only a third of the iron atoms involved in reordering (see figure 2.10). Once again, the models yield results which are generally correct, surrounding the experimental curve. In this example, once again, the QC model seems to be the closest to reality.

2.4.5.2. *Determination of the enthalpy of transformation*

In the two models, we have found the expression of the Helmholtz energy given by the two relations [2.120] and [2.146]. In each case, we can calculate the internal energy of configuration (which is the same as the enthalpy, as the pressure has no influence at all) using the relation:

$$U = T^2 \frac{\partial(F/T)}{\partial T} \qquad [2.150]$$

In general, it is preferable to calculate the difference $\frac{U_T - U_{T=0}}{N\mathrm{k_B}T}$ or the difference $\frac{U_T - U_{T=\infty}}{N\mathrm{k_B}T}$. Note that we have:

$$U_T - U_{T=\infty} = U(s) - U(0) \qquad [2.151a]$$

and $U_T - U_{T=0} = U(s) - U(1)$ [2.151b]

We find the following results:

For the GBW model:

$$U(s) = U(0) + \frac{Ns^2 w_{\mathrm{AB}}}{2} = U(1) - \frac{N\left(1-s^2\right) w_{\mathrm{AB}}}{4} \qquad [2.152]$$

For the QC model, in light of relation [2.131]:

$$U(s) = < E(s) > = \frac{Nz}{2}\varepsilon_{\mathrm{AB}} - \frac{Nw_{\mathrm{AB}}}{2}\frac{\left(1-s^2\right)}{\beta_s + 1} \qquad [2.153]$$

Source	GBW	QC	Experimental
$\frac{U_{T_c} - U_{T=0}}{N\mathrm{k_B}T_c}$	0.50	0.49	0.43

Table 2.6. *Comparison between experimental results and models of the values of the internal energy for the alloy CuZn (calculations performed with z = 8), at the critical temperature 742 K*

On the basis of these expressions, we can perform the calculation at any temperature. For example, the calculation for the alloy CuZn, at the critical temperature of $T_c = 742$ K with $z = 8$, gives us the results collected in Table 2.6. Again, we see that both models yield results that are fairly similar to each other, and also values that are not too different from the experimental reality.

2.4.5.3. *Measurement of the critical temperature and its variations with composition*

We have performed the calculation of the critical temperature in the two models in the case of the compositions $x = y = ½$. As pointed out, it is entirely possible to perform the same calculations for a fraction of site $y = ½$ and a molar fraction of one of the components x between 0 and 1. Figures 2.17(a) and 2.17(b), respectively, for the alloy CuZn and Fe_3Al, give the variations of the critical temperature as a function of the composition.

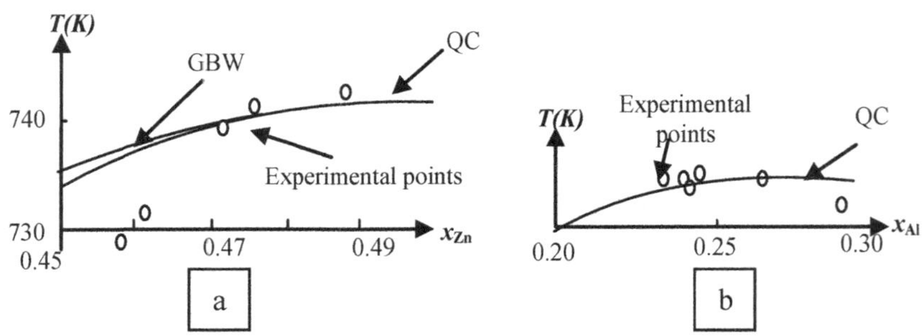

Figure 2.17. *Critical temperature as a function of the composition. Comparisons between models and experience. a) Case of CuZn and b) case of Fe₃Al (according to [SYK 37])*

In Figure 2.17(a), the results given by the two models are very similar, and differ notably from the experimental results when the composition deviates from $x = 1/2$. In Figure 2.17(b), the QC model gives correct results around the composition Fe_3Al.

2.4.5.4. *Variations of specific heat capacity with temperature*

The variations in the specific heat capacities with temperature are determined on the basis of the measured or calculated values of the internal energy using the relation:

$$C(T) = \frac{\partial U(T)}{\partial T} = \frac{\partial U(s)}{\partial T} \qquad [2.154]$$

Figure 2.18 shows the variations of the average of the specific heat capacities for copper and zinc (dotted curve), and an experimental curve for the CuZn.

The curve derived from a model results from the sum of the configurational specific heat capacity calculated by that model and the specific heat capacities of the pure metals, in order to compare the result with the experimental value, which gives the total specific heat capacity.

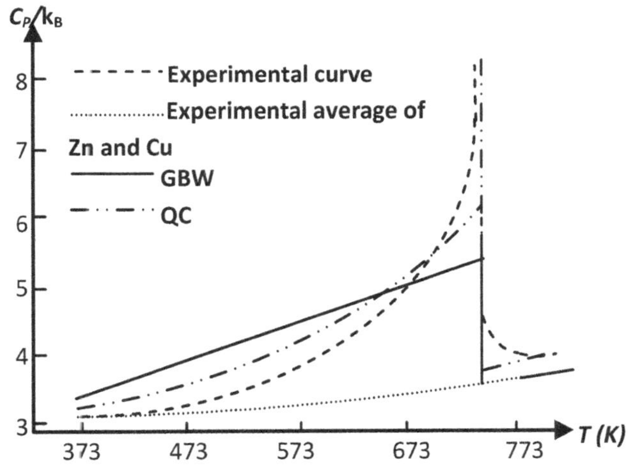

Figure 2.18. *Specific heat capacity (per atom) of CuZn – comparison of the models and the experimental results [SYK 37]*

We can see that the comparison is very poor – particularly in the vicinity of the critical temperature, for which both models are absolutely incapable of accounting for the discontinuity suggested by the experimental curve – a discontinuity which enables us to deduce that in the case of the copper–zinc alloy, the order–disorder transformation is a second-order transformation.

Ultimately, the comparison of the models with the experimental reality gives qualitative agreements for most experiences but the comparison is not at all good in terms of the specific heat capacities.

The differences which we observe need to be viewed in relation to the hypotheses common to the two models – first an interatomic distance and therefore an energy of interaction which is independent of the degree of order. More recently, Koslov and Ginsburg attempted to introduce a correction to the energy of interaction, expressed by a correlation between the energy and the square of the differences between an average interatomic distance and the true distance. They applied this correlation to the GBW model. At the most, they managed to obtain a curve giving the variation of the degree of order as a function of the temperature, close to that obtained directly by the QC model as represented in Figure 2.14(a).

It seems that the main shortcomings of the simple models we have presented here are due essentially to the hypothesis of interactions limited to that which exists between two closely neighboring atoms.

2.5. Determination of the activity of a component of a solid solution

We will examine the methods for determining the activities, or the activity coefficients, of the components of a solid solution whose composition is known. Note in passing that this determination of the activity can quickly lead us to the chemical potential of that component, by knowledge of the chemical potential of that component in the reference state.

Remember that the activity and the activity coefficient of a component in a solution depend both on the composition of the solution and its temperature. Therefore, a determined value of the activity of a component is only valid at a given temperature and known composition of the solution.

It must also be remembered that there are several activities for the same component, depending on the convention chosen to define it – i.e.:

– convention (I): pure-substance references;

– convention (II): infinitely dilute solution reference;

– convention (III): reference of a molar solution of all solutes.

Therefore, if it is not generally applicable, each method must specify the convention in which the determination is performed.

The methods to determine the activities (or the activity coefficients) are divided into two main categories:

– experimental methods, which are dealt with in this chapter;

– methods using a model of solution, methods which arise directly from the study of the different models (see section 2.3).

The experimental methods which we are about to examine are based on the properties of solid solutions, and more specifically, on their behavior in a system at physico-chemical equilibrium. The component whose activity we wish to measure is involved in a state of physical, chemical or electrochemical equilibrium and the only unknown degree of activity is that which we are searching for.

Numerous methods which are applicable both to solid solutions and liquid solutions will not be examined in detail but simply mentioned. A detailed description of these methods is given in Chapter 5 of Volume 2 in this collection, devoted to liquid phases (see Bibliography). Alongside these general methods, there are a few that are more specific to solid solutions. The experimental difficulty which is often encountered with solid solutions arises from the fact that the measurements need to be carried out at high temperature, which sometimes complicates display technology.

2.5.1. *Methods common to solid solutions and liquid solutions*

These methods are divided into two categories: computational methods and experimental methods.

2.5.1.1. *Calculating an activity coefficient on the basis of the known values of other coefficients*

Let us recall the two methods which are, in fact, computational methods based on the knowledge either of the activity of the other species in the solution or of the activity of the species sought at a temperature other than the desired one.

2.5.1.1.1. Calculating the activity of a component when we know those of the other components of the solution

This method is based on the integration of the Gibbs–Duhem relation at fixed temperature and pressure.

$$\sum_{i=1}^{N} x_i \, \mathrm{d}\ln\gamma_i = 0 \qquad [2.155]$$

We suppose that, for all the compositions between a known state (usually the reference state) and the composition under study, we know the activities (or the activity coefficients) of all the components of a solid solution in the same convention, except for one between them, and we need to calculate the unknown activity of that component for the chosen composition. This method is valid regardless of the convention adopted. Thus, we obtain the activity (or the activity coefficient) in the chosen convention for the known values.

To determine the activity coefficient γ_j of the component j whose molar fraction is x_j, we write, in light of relation [2.155]:

$$\mathrm{d}\ln\gamma_j = -\sum_{k\neq j} \frac{x_k}{x_j} \mathrm{d}\ln\gamma_k \qquad [2.156]$$

This relation is integrated between two states p and q. Each state is characterized by a composition of the solution:

$$\int_p^q \mathrm{d}\ln\gamma_j = -\sum_{k=j}\int_p^q \frac{x_k}{x_j}\mathrm{d}\ln\gamma_k \tag{2.157}$$

Generally, as the lower bound p, we choose the reference state, which gives a value of 1 for the activity coefficient γ_j, and as the upper bound q we choose the solution in which we are looking for the coefficient. Usually, we operate by numerical integration.

2.5.1.1.2. Determination of the activity of a component at one temperature if we know its activity at a different temperature

This method is based on Helmholtz's second relation. Let us, respectively, write $\overline{H_i^*}$ and $\overline{H_i^x}$ for the partial molar enthalpies of the component i in the reference solution and in the solution at hand, for which the molar fraction of the component i is x. We have:

$$\frac{\partial(\mu_i/T)}{\partial T} = -\frac{\overline{H_i^x} - \overline{H_i^*}}{T^2} \tag{2.158}$$

On this relation, we superimpose the expression of the derivative of the logarithm of the activity coefficient with the temperature:

$$\mathrm{R}\frac{\partial\ln\gamma_i^{(*)}}{\partial T} = \frac{\partial(\mu_i/T)}{\partial T} \tag{2.159}$$

From this, we deduce the activity coefficient of the component i at temperature T', for the molar fraction x:

$$\frac{\partial\ln\gamma_i^{(*)}}{\partial T} = \frac{\overline{H_i^*} - \overline{H_i^x}}{\mathrm{R}T^2} \tag{2.160}$$

It is possible to integrate this expression, in the chosen convention, if we know the variations of the enthalpies with the temperature. We obtain the activity coefficient of the component i at temperature T if we know the values of the activity coefficient of the same component at another temperature $\gamma_{i(T')}^{(*)}$.

Usually, these enthalpies are considered to be constant in too great a temperature interval, so between two temperatures T and T', we can integrate relation [2.160] in any one of the references (denoted by *) in the form:

$$\ln \gamma_{i(T)}^{(*)} - \ln \gamma_{i(T')}^{(*)} = \frac{\overline{H_i^*} - \overline{H_i^x}}{\mathrm{R}}\left(\frac{1}{T'} - \frac{1}{T}\right) \qquad [2.161]$$

The partial molar enthalpies $\overline{H_i^*}$ and $\overline{H_i^*}$ are then constant.

2.5.1.2. *Determination of the activity on the basis of the measured value of the equilibrium pressure*

The methods using the vapor tensions derive measurements directly from the properties of the equilibrium between the solid solution and one of its gaseous components, which is a consequence of the equality of the chemical potentials of the component in question in the solution and in the gaseous phase at equilibrium with it. In general, the measurements take place at sufficiently high temperatures and sufficiently low pressures so that the gaseous phase can be considered to be perfect in all senses of the word. This equality gives us the very general relation between the partial pressure of the component i, its activity $a_i^{(S)}$ in the solid solution and an equilibrium constant $K_i^{(SV)}$:

$$\frac{P}{a_i^{(S)}} = K_i^{(SV)} \qquad [2.162]$$

Experimentally, these methods only require the measurement of the partial pressure, with a mass spectrometer, and the measurement of the composition of the solid phase. The best technique is thermogravimetry. This method is primarily used in the case of dissolution of gases, because then only that gas is present in the gaseous phase. The partial pressure can then be taken to be equal to the total pressure if that gas is pure. We then proceed as follows: we set a low value P_1 of the pressure P_i and wait for equilibrium to be achieved, and then take away the mass m_1. We then raise the pressure successively to values of P_2, P_3, etc. Each time, we wait for equilibrium to be reached and take away the masses m_2, m_3, etc. Finally, we create a vacuum above the sample and measure the mass m_0, which will be the initial mass of the solution: in fact, the mass of the non-volatile solvent on its own. From these measurements, we deduce the table of the couples of pressure–composition in the solution.

We can distinguish two approaches:

– one, called *the direct method*, which is mainly used with a pure-substance reference convention – i.e. for a single-component solution for which we choose convention (I);

– the other, which is based on the measured value of Henry's constant, which is mainly used for solutes with the choice of convention (II).

We will base our examination on the simple case where only one component of the solution is present in the gaseous phase, in which it is pure.

2.5.1.2.1. Measurement by the direct method

For this method, we choose to adopt convention (I), the solid-vapor equilibrium constant, with the saturating vapor pressure of the pure solute under study P_i^0, which is expressed by:

$$\frac{P}{\gamma_i^{(I)} x_i^{(S)}} = P_i^0 \qquad [2.163]$$

We need to be very careful when it comes to the choice of the pressure P_i^0. Indeed, it is the saturating vapor pressure of the component i in the same state of aggregation as the solution. This means that the pure substance must be considered to be crystallized in the same crystalline system as the solution. However, the solute is usually not solid if it is pure in the conditions in which the solution is studied. Thus, to begin with, we can calculate this vapor pressure on the basis of the known value – usually for a much lower temperature. Yet, even at that temperature where the component i is solid, it generally does not crystallize in the same crystalline system as the solid solution. Therefore, we need to simulate a change of solid phase for which we do not have either the latent heat or the associated entropy. We can obtain a correct approximation by ignoring the variation of entropy between two solid states and choosing a variation in enthalpy of around 1/4 of the latent heat of fusion of the same pure component.

Relation [2.163] then directly gives us:

$$\gamma_i^{(I)} = \frac{P}{x_i^{(S)} P_i^0} \qquad [2.164]$$

Thus, if we know the pressure–composition couples, we can calculate the activity coefficient for the different compositions at the measuring temperature.

2.5.1.2.2. Method using Henry's constant

We remain in the case of the presence of a single component in the gaseous phase. If we adopt convention (II) for a solute, the expression of the solid-vapor equilibrium for the component i is written, at a given temperature, in the form:

$$\frac{P_i}{\gamma_i^{(S)(II)} x_i^{(S)}} = K_i^{(SV)} \qquad [2.165]$$

If the gaseous mixture can be considered to be perfect, the equilibrium constant is the product of the saturating vapor pressure of the pure component with Henry's constant (K_{iH}) for the same component i and, for the equilibrium state, relation [2.165] gives us:

$$\frac{P_i}{\gamma_i^{(S)(II)} x_i^{(S)}} = K_i^{(LV)} = P_i^0 K_{iH} \quad [2.166]$$

If the solution were perfect, the solute i would obey Henry's law, and thus we would have the equation:

$$\frac{P_i}{x_i^{(S)}} = P_i^0 K_{iH} \quad [2.167]$$

This means that the constant K_{iH} is the limit, when the content $x_i^{(S)}$ tends toward zero, i.e. when $\gamma_i^{(S)(II)}$ is equal to 1, of the ratio $\frac{P_i}{P_i^0 x_i^{(S)}}$, which is:

$$K_{iH} = \lim_{x_i^{(S)} \to 0} \frac{P_i}{P_i^0 x_i^{(S)}} \quad [2.168]$$

We determine the value of the ratio $\frac{P_i}{P_i^0 x_i^{(S)}}$ for various decreasing values of $x_i^{(L)}$, we extrapolate the curve obtained at $x_i^{(S)} = 0$. The order then gives us the constant K_{iH}.

The extrapolation is shown in Figure 2.19. The curve must have a limit with a horizontal slope. If this is not the case, the fact is that the measurements of the ratio $\frac{P_i}{P_I^0 x_i^{(S)}}$ have not been carried out for sufficiently low values of the molar fraction of i in the solid and thus achieve the validity of the limit law with a correct degree of precision.

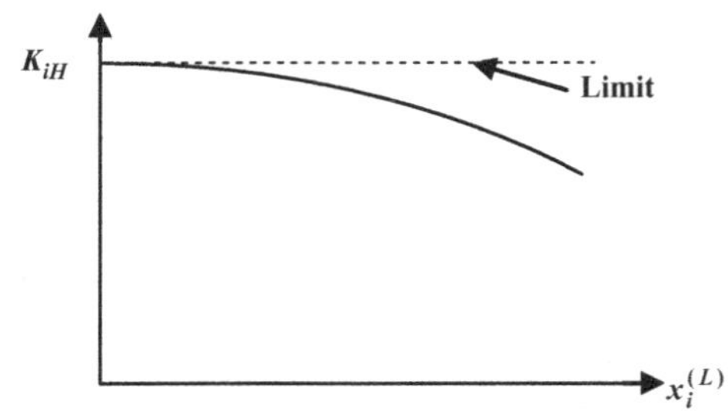

Figure 2.19. *Determination of Henry's constant*

We then feed that extrapolated value back into equation [2.166] and deduce the activity coefficient of the component i in convention (II):

$$\gamma_i^{(S)(II)} = \frac{P_i}{x_i^{(S)} P_i^0 K_{iH}} \qquad [2.169]$$

As before, this law requires us to know the appropriate value of the vapor pressure P_i^0.

NOTE.– The expressions which we have used suppose that the solute and the gas have the same molecular form. If, on the other hand, the solute were in monatomic form and the gas in diatomic form, all the pressures which come into play in expressions [2.164] and [2.169] would be replaced by square roots, expressing the dissociation of the gas upon dissolution.

2.5.2. *Methods specific to solid solutions*

We will now present two methods for measuring the activities which are more specifically designed for solid solutions. The first method, involving the solid solution in a reaction which does not involve any other solution. Note that this method could be applied to liquid solutions but usable reactions are then very rare. The second method is an electrochemical method designed for single-phase metal alloys.

2.5.2.1. *Determination of the activity of a component on the basis of the composition of a system at chemical equilibrium*

We can determine the activity of a component of a solution by involving it in a chemical equilibrium. The method is particularly interesting when the compound under study is the only component in solution which participates in the chosen chemical equilibrium.

In order to illustrate the method, we describe how to determine the activity coefficient of carbon in a steel. Remember that a steel is a solid solution of carbon in iron. We will use this steel in the form of sheets, with which we will seek to realize the equilibrium:

$$CO_2 + C = 2CO \qquad [2R.3]$$

At equilibrium, we measure the ratio P_{CO}^2 / P_{CO_2} for different levels of carbon content. The application of the law of mass action in reference (II) is written as:

$$K^{(II)} = \frac{P_{CO}^2}{P_{CO_2} \gamma_c^{(II)} x_c} \qquad [2.170]$$

In the chosen convention, the activity coefficient tends toward 1 if the carbon content tends toward zero. Therefore, we can write:

$$\lim_{x_c \to 0} \frac{P_{CO}^2}{P_{CO_2} x_c} = K^{(II)} \qquad [2.171]$$

Thus, we take the ratio $P_{CO}^2 / x_c P_{CO_2}$ as a function of the molar fraction of carbon x_c (Figure 2.20). Extrapolation of the curve for $x_c = 0$ gives the value of the equilibrium constant $K^{(II)}$. Thus, to calculate the activity coefficient at the content level chosen, we apply the relation:

$$\gamma_c^{(II)} = \frac{P_{CO}^2}{P_{CO_2} x_c K^{(II)}} \qquad [2.172]$$

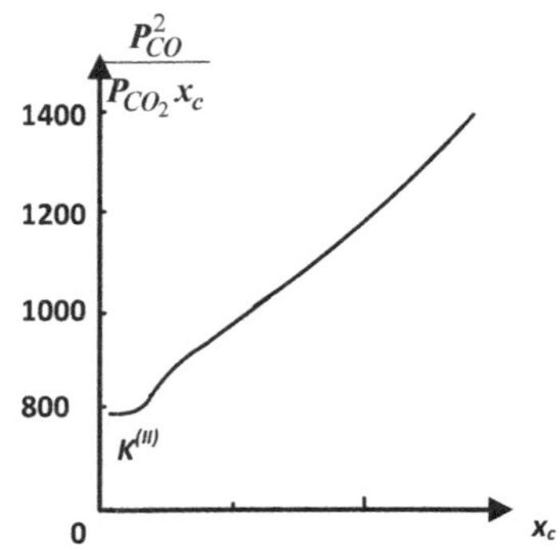

Figure 2.20. *Determination of the equilibrium constant between carbon, CO and CO2*

NOTE.– Certain authors propose to determine the activity coefficient of carbon in the pure-substance reference (I) using the relation:

$$\gamma_c^{(I)} = \frac{P_{CO}^2}{P_{CO_2} x_c K^{(I)}} \quad [2.173]$$

In order to do so, as an equilibrium constant $K^{(I)}$, they use that determined by achieving the equilibrium [2R.3] with pure carbon in graphic form. Thus, they make the approximation of neglecting the Gibbs energy of the passage of pure carbon from the graphic state to a hypothetical state of a carbon crystallized in the same system as the iron in steel. It is true that this Gibbs energy may not be hugely different from that of the passage from the graphite form to the diamond form, which could, therefore, be used as a best approximation.

2.5.2.2. *Measuring the activity of a component non-ionic conductive solution (metal solution)*

If we consider an electrode reaction which is written as:

$$\sum_i \mathrm{v_i M_i} + \mathrm{v_e e} = 0 \quad [2R.4]$$

The potential is given by the Nernst equation.

$$e = e^0 + \frac{RT}{\nu_e F} \ln \prod_i |M_i|^{\nu_i} \qquad [2.174]$$

In this expression, we see the activities of all the species M_i involved in the equilibrium [2R.4].

In general, the electrode is made of a pure metal whose activity is 1. However, if the electrode is made of an alloy forming a solution, relation [2.174] will include the activity of the active metal in the alloy. However, relation [2.174] contains several other activities which are usually unknown. Let us examine the problem on a concrete example.

Thus, we will envisage measuring the activity of lead in a silver–lead alloy of the molar fraction x_1 of lead. The electrode is composed of the alloy under examination and a lead salt, so, for example:

$$\text{Pb (solid solution) } (x_2) \text{ / } Pb(CH_3COO)_2 \text{ (aqueous solution)}$$

The potential of this electrode is written as:

$$e = e^0 + \frac{RT}{2F} \frac{|Pb^{++}|}{a_1} \qquad [2.175]$$

The activity a_1 of the lead can only be deduced from the measured value of the potential of that electrode if the activity of the lead in the aqueous solution is known.

In order to get around this difficulty, we will construct a cell, similar to concentration cells, but the dissymmetry of the cell will be assured by a difference in concentration between the two electrode alloys, and not by two ionic compartments.

We will include our electrode in a cell whose other electrode is the same alloy but with a different molar fraction of lead x_2. The two electrodes are immersed in a solution of a lead salt – e.g. acetate. Thus, we will create the following electrode:

$$\text{Pb (solid solution) } (x_2) \,/\, \text{Pb}\left(\text{CH}_3\text{COO}\right)_2$$
$$\text{CH}_3\text{COOH } \,/\, \text{Pb (solid solution) } (x_1)$$

For the two electrodes, the reference state of lead is the same, so the electromotive force of the battery is written as:

$$E = \frac{\text{R}T}{2\text{F}} \ln \frac{a_1}{a_2} \qquad [2.176]$$

By separating the molar fraction and the activity of lead in our solid solution under study, this expression can be put in the form:

$$\frac{2\text{F}E}{\text{R}T} + \ln x_1 = \ln a_2 - \ln \gamma_1 \qquad [2.177]$$

When the molar fraction of lead in our solution x_1 tends toward zero, the activity coefficient $\gamma_1^{(II)}$ tends toward 1, so we can write:

$$\lim_{x_1 \to 0} \left(\frac{2\text{F}E}{\text{R}T} + \ln x_1 \right) = \ln a_2^{(II)} \qquad [2.178]$$

To use this last relation, we can measure the electromotive forces of the cell for different values of the molar fraction x_1 and plot the curve giving the value $y = \frac{2\text{F}E}{\text{R}T} + \ln x_1$ as a function of the molar fraction x_1. Figure 2.21 shows such a curve.

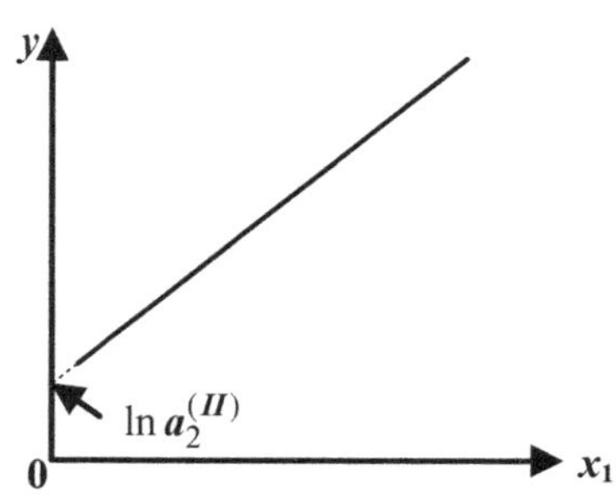

Figure 2.21. *Obtaining the activity of an element of an alloy*

The extrapolation of this curve to the origin of the abscissa axis enables us to calculate the value of $\ln a_2^{(II)}$, which is the activity of lead in the second compartment. We can then easily calculate the activity coefficient sought for each concentration by using relation [2.177] in the form:

$$\ln \gamma_1^{(II)} = -\frac{2\mathsf{F}E}{\mathrm{R}T} - \ln x_1 + \ln a_2^{(II)} \qquad [2.179]$$

The method is particularly elegant, and does not require the use of a junction.

Another very similar method can also be used – particularly to measure the activity of an element in a metal alloy at high temperature. We create a cell with two electrodes, one of which is the pure metal and the other is the solid solution, using a solid ionic conductor as an electrolyte. By measuring the electromotive force of the cell, at the chosen temperature, we will be able to calculate the activity of an element of the alloy at the desired temperature.

We will illustrate the method on the following example.

Thus, let us consider an alloy of tantalum and molybdenum comprising a solid solution. We wish to determine the activity of the tantalum in that solution at 1300 K. For this, we use a solid electrolyte with ionic conduction, which is a solid solution of thorium oxide ThO_2, containing 6% mass of yttrium oxide (Y_2O_3). We construct the following cell:

Ta (pure metal $+ Ta_2O_5$ compact) / ThO_2 (6%Y_2 O_3)/

Ta (solid solution $(x_2) + Ta_2O_5$ compact)

The reactions at the electrodes are:

– on the side of pure tantalum:<Ta>→Ta^{+5}+5e;

– on the side of the alloy: Ta^{+5}+5e →<<Ta>>.

The electromotive force of the cell has been measured by Singhal and Worrel at 1200 K for different proportions x_2 of tantalum in the alloy.

The cell reaction is simply the passage of tantalum atoms from the pure metal into the alloy. The variation of the Gibbs energy with this reaction is:

$$\Delta_s G = \bar{G}_2 - g_2^0 = \mathrm{R}T \ln a_2^{(I)} \qquad [2.180]$$

However, this Gibbs energy is linked to the electromotive force of the cell by the relation:

$$E = -\frac{\Delta_s G}{5\mathrm{F}} = \frac{\mathrm{R}T \ln a_2^{(I)}}{5\mathrm{F}} \qquad [2.181]$$

From this, we deduce the activity of the tantalum in the alloy:

$$a_2^{(I)} = \exp - \frac{5\mathrm{F}E}{\mathrm{R}T} \qquad [2.182]$$

The difficulty in using such a method is threefold. First, we need to work at a high temperature; second, we need to ensure very good contact between the solids; and finally, we need to make sure that the electrolyte works well for ionic conduction.

In Chapter 4, we will return to solid solutions, after having introduced structural elements and non-stoichiometry in Chapter 3. This will enable us to correlate the notion of a solid solution with that of non-stoichiometry.

3

Non-stoichiometry in Solids

The concept of a structure element is absolutely crucial in studying the thermodynamics of non-stoichiometry in solids.

In most crystalline mineral solids, the concept of a molecule has no physical reality. For example, if we look at zinc oxide, whose formula is ZnO, we cannot attribute a specific oxygen atom to a particular zinc atom.

Additionally, in comparison to an ideal crystal, comprising zinc ions and oxygen ions regularly arranged in space, many solids exhibit defects such as the presence of a zinc ion in an abnormal position, qualified as interstitial, because it is between the positions of the ions in the ideal crystal.

Moreover, many of these compounds exhibit differences from stoichiometry, meaning that they do not rigorously obey the chemical formula of the ideal compound, such as zinc oxide which, in reality, presents an excess of zinc in relation to oxygen (which is known as overstoichiometry in zinc). Strictly speaking, its formula should be written as $Zn_{1+x}O$, with the value of x varying under the influence of various stresses (oxygen pressure, temperature, etc.).

3.1. Structure elements of a solid

Because of all these peculiarities, the description of a solid simply through the lens of its atoms, ions or molecules cannot give an account of numerous physical and physicochemical properties, and that description must be precise, involving the nature of the sites occupied by the atoms – i.e. specifying their environment. This is the description of the solid in its *structure elements*.

3.1.1. *Definition*

A structure element is an atom, an ion or a vacancy (an empty space) at a specific site in the crystal. The concept of a structure element brings together that of a chemical species and that of its environment by the nature of the crystalline site in question. A structure element could be:

– an atom or an ion of the solid at a normal site (in the sense of the ideal solid);

– an atom or an ion of the solid at an abnormal site (atom substituting an atom of a different type, atom in an interstitial position, etc.);

– a vacancy in the normal (an empty space in the ideal solid);

– an empty interstitial position which we can consider to be a vacancy in an interstitial position;

– a foreign atom or ion in the solid at a specific site (an example is offered by chromium in place of zinc in zinc oxide).

Thus, we can completely describe the solid by enumerating its structure elements, assimilating the free electrons and the electron holes likely to be present to structure elements, for the sake of completeness.

The structure elements can carry different charges. The "effective charge" of a structure element, denoted by q_e, is the difference between its real charge, represented by q_r, and the charge that the structure element occupying the same site in the ideal crystal would have, or the "normal charge", denoted by q_n:

$$q_e = q_r - q_n \quad [3.1]$$

A structure element is said to be non-ionized if its effective charge is zero. Consequently, an ionized structure element may perfectly well be a neutral atom.

The normal structure elements of the solid – i.e. the elements that are present in the ideal solid (an atom or ion in the solid at a normal site or an empty interstitial position) – have an effective charge of zero.

Structure elements other than the normal elements of the solid are often referred to as the solid's *point defects*.

3.1.2. *Symbolic representation of structure elements*

In the same way that in chemistry, we see the need to symbolically represent the atoms and molecules, it has become necessary to have a symbolic representation of the structure elements. This notation system must provide three types of information about an element:

– the atom or chemical element in question;

– the site on the lattice that is occupied (in reference to the ideal solid);

– the effective charge (or the real charge). We tend more to choose the effective charge, because it is zero for all normal elements, which are generally present in very large numbers.

The International Union of Pure and Applied Chemistry (IUPAC) prescribes the use of Kröger's notation. Tables 3.1, 3.2, 3.3, 3.4 and 3.5 all show this notation for the different types of structure elements, applied to the (fictitious) example of alumina.

The symbol ' above and to the right of an element means that the element has an effective charge of –1; the symbol ° represents the effective charge +1; the effective charge of 0 is represented by the superscript x, but it is not obligatory to use this sign at all. An element in an interstitial position is denoted by the subscript symbol "i".

Al^{3+} ion in the normal position (q_e= 0)……..Al_{Al}
O^{2-} ion in the normal position (q_e= 0)………..O_O
Empty interstitial (q_e= 0)……………………..V_i

Table 3.1. *Symbols of normal structure elements*

Free electron (q_e= -1)…………e'
Free electron holes (q_e= 1)…….h°

Table 3.2. *Symbols of the free charges*

Note that a vacancy is denoted by the symbol V. In view of the possibility of confusion with the representation of vanadium, which has the same symbol, the IUPAC advocates that, where structure elements are involved, the symbol for vanadium is replaced by "Va".

Al^{3+} ion in an interstitial position (q_e= 3) ….. $Al_i^{\circ\circ\circ}$
O^{2-} ion in an interstitial position (q_e= -2)…….. $O_i^{''}$
Al atom in an interstitial position (q_e= 0)……..Al_i

Table 3.3. *Symbols for structure elements in interstitial positions*

Vacancy of an Al^{3+} ion (q_e= -3)………………………………. $V_{Al}^{'''}$
Vacancy of an O^{2-} ion (q_e= -2) ……………………………….. $V_O^{\circ\circ}$
Vacancy of an Al^{3+} ion trapping an electron hole (q_e= -2)……. $V_{Al}^{''}$
Vacancy of an O^{2-} ion trapping an electron (q_e= 1)…………… V_O°

Table 3.4. *Symbols for vacancy structure elements*

Ion S^{2-} in substitution of an O^{2-} ion (q_e= 0)…………..S_O
Ion Mg^{2+} in substitution of an Al^{3+} ion (q_e= -1)… $Mg_{Al}^{'}$
Ion Zr^{4+} in substitution of an Al^{3+} ion(q_e= 1)…….. Zr_{Al}°
Ion Li^{+} in an interstitial position (q_e= 1)…………. Li_i°

Table 3.5. *Symbols for structure elements occupied by foreign elements*

Associations between structure elements are represented by indicating, in parentheses, the associated elements without their effective charge and, after the parenthesis, the effective charge of the whole thing. For example, the association, in alumina, of an aluminum ion vacancy and an oxygen ion vacancy is written as: $\left(V_{Al}V_O\right)'$. Note that we would obtain the same entity and therefore the same notation for the effective charge, -1, if each of those vacancies were ionized, having respectively trapped an electron hole and an electron.

3.1.3. *Building unit of a solid*

The *building unit* (or building block) of a solid is a combination of structure elements, such that the addition or subtraction of such a combination does not alter the ratios between the numbers of the various sites in the crystal (conservation of the structure). The true solid can only be constructed by the juxtaposition of such units, in varying numbers; for this reason, we sometimes call these entities the *building blocks* of the crystalline solid.

For a compound AB (with one interstitial position per site of A), the building units may be the sum of appropriate structure elements, as in the following examples:

$$A_A + B_B + V_i\,;\ V_A + V_B + V_i\,;\ F_A + B_B + V_i$$

where F is the symbol for a foreign atom other than A and B.

Differences such as $V_B - B_B$ or $B_i - V_i$ also constitute building units.

3.1.4. *Description and composition of a solid*

By introducing the concept of structure elements, we are able to consider them as the components of a solid solution because, in a real solid, their proportions are likely to vary continually (the compositional proportion of a structure element can vary without phase change, which is the distinguishing feature of a component in a solution). We will see that we can take the solid described in terms of structure elements and apply the concepts of thermodynamics of solutions to it. In order to do so, we need to define properties which quantify the composition of the solid phase in terms of each of the structure elements which make it up. Several types of properties are used for this purpose.

3.1.4.1. *Fraction of sites*

The fraction of sites is the number of structure elements of a given type, divided by the total number of sites of that type of structure elements (whether occupied or vacant) in the same volume of the solid.

Take, as an example, the case of barium oxide which, as we will see, has barium cations in interstitial positions. The fraction of sites for those barium ions is given by the ratio of the quantity (number of moles) of barium ions in that interstitial position to the total quantity of interstitial positions, whether occupied or otherwise, in that same volume. Thus, we will represent that fraction of sites as:

$$\frac{n_{\mathrm{Ba}_i^{\circ\circ}}}{n_{\mathrm{V}_i} + n_{\mathrm{Ba}_i^{\circ\circ}}} \quad [3.2]$$

In cases where the structure element in question is very dilute, this expression can be simplified to:

$$\frac{n_{\mathrm{Ba}_i^{\circ\circ}}}{n_{\mathrm{V}_i}} \quad [3.3]$$

It is this value that we will use in our thermodynamic studies using structure elements.

3.1.4.2. *Concentration*

To define the concentration, we use the same definition as that given for liquid solutions, i.e. it is the quantity of the structure element in question, divided by the volume of the phase. In our example of barium cations in interstitial positions in barium oxide, the concentration is given by:

$$C = \frac{n_{\mathrm{Ba}_i^{\circ\circ}}}{V} \qquad [3.4]$$

3.1.4.3. *Atomic fraction*

The atomic fraction is the quantity of the structure element divided by the total quantity of matter contained in the volume. In our example, the atomic fraction of interstitial barium ions would be:

$$\frac{n_{\mathrm{Ba}_i^{\circ\circ}}}{n_{\mathrm{Ba}} + n_{\mathrm{O}}} \qquad [3.5]$$

where n_{Ba} and n_{O}, respectively, represent the total quantities of barium and oxygen.

It is easy to move from one of the properties defined above to another, using the molar masses and densities.

3.2. Quasi-chemical reactions in solids

The calculations for equilibrium states involving structure elements will be carried out using the quasi-chemical method, with equilibrium between reactions creating or involving those structure elements.

3.2.1. *Definition and characteristics of a quasi-chemical reaction between structure elements*

When a solid enters into a reaction, we accept that it is, in fact, the structure elements which react. Similarly, the production or elimination of a defect in a solid is the consequence of reactions which, necessarily, involve those structure elements. All these reactions can be represented by a formulation, similar to the formulation of conventional chemical reactions, but using the symbol system of structure elements. We obtain what are known as quasi-chemical reactions.

Such reactions have a certain number of characteristics:

– they must not alter the structure of the solid, and therefore must preserve the ratios between the numbers of the different sites in the solid. One consequence of this property is that a quasi-chemical reaction must be able to be written by using building units (we will exploit this property in section 3.3 to define the chemical potentials of the structure elements), or the structure elements, as a reaction in an aqueous solution can be written either using the ions or the neutral molecules that surround those ions;

– quasi-chemical reactions must preserve the electrical charges, and in particular the effective charges;

– they must preserve the chemical elements.

As an example, we will examine the reaction between gaseous oxygen and barium oxide. This oxide has interstitial cations and contains free electrons and therefore an excess of cations. At equilibrium, this overstoichiometry is greater or lesser depending on the oxygen pressure, which reigns over the solid. This is due to a reaction which it is tempting to write with the usual chemical symbols in the form:

$$\mathrm{Ba}_{(1+x)}\mathrm{O} + \frac{x}{2}\mathrm{O}_2 = (1+x)\mathrm{BaO}$$

where $\mathrm{Ba}_{(1+x)}\mathrm{O}$ denotes the non-stoichiometric form of barium oxide $(x > 0)$.

Writing the reaction this way exhibits several disadvantages:

– it suggests that, under the influence of the oxygen, the barium oxide automatically becomes stoichiometric, and thus that there are two forms of barium oxide – one stoichiometric and the other non-stoichiometric;

– it does not show which barium ions are directly involved in the reaction, because it does not distinguish between the ions in a normal position and those in an interstitial position;

– it involves stoichiometric numbers (x) that vary with the pressure, which is rather odd for a stoichiometric number.

Thus, we will systematically reject this type of formulation, which would also exhibit other disadvantages when we attempt to apply the laws of equilibrium, such as the law of mass action (it is difficult to imagine the role of x in such a scenario).

To write the quasi-chemical reaction, we proceed as follows:

– we write the ideal stoichiometric oxide in structure elements. We have the equivalence:

$$\mathrm{BaO} \equiv \mathrm{Ba_{Ba}} + \mathrm{O_O} + \mathrm{V_i}$$

– we do the same for the non-stoichiometric form:

$$\mathrm{Ba}_{(1+x)}\mathrm{O} \equiv \mathrm{Ba_{Ba}} + \mathrm{O_O} + x\mathrm{Ba_i} + (1-x)\mathrm{V_i} + 2xe'$$

– we now substitute, into the above chemical equilibrium, the two forms of oxide as written above. After simplification of the elements present on both sides of the equation, we obtain the quasi-chemical reaction:

$$\mathrm{Ba_i^{\circ\circ}} + 2e' + \tfrac{1}{2}\mathrm{O_2} = \mathrm{Ba_{Ba}} + \mathrm{O_O} + \mathrm{V_i}$$

This clearly shows us that the addition of gaseous oxygen takes place at the expense of the barium ions in interstitial position and the free electrons, and only them. Note that the quantity x has disappeared, and we see that increasing the oxygen pressure, at equilibrium, decreases the number of interstitial barium ions (shifting the equilibrium toward the right), but this does not mean that the oxide automatically becomes stoichiometric. Note that this formulation preserves the ratios between sites, the effective charges and the elements, but the reaction results in the appearance of a new building unit, causing an increase in the dimensions of the crystal.

Quasi-chemical reactions may take place in the homogeneous phase within the solid, or be heterogeneous and occur at an interface between the solid and another phase. We will examine the different types of quasi-chemical reactions.

3.2.2. *Homogeneous quasi-chemical reactions in the solid phase*

These reactions take place within the solid phase. We can distinguish four categories of homogeneous reactions: electron reaction, reactions creating disorders (and conversely annihilating disorders), ionization reactions and addition reactions.

3.2.2.1. *Electron reaction*

An electron reaction is the recombination of the free electrons and electron holes, which is written as:

$$e' + h^{\circ} = 0 \qquad [3R.1]$$

The electrons involved are those in the conduction band, and the holes are at the top of the valence band in the energy diagram (see Figure 3.1). The Gibbs energy associated with that reaction is represented by the height of the gap.

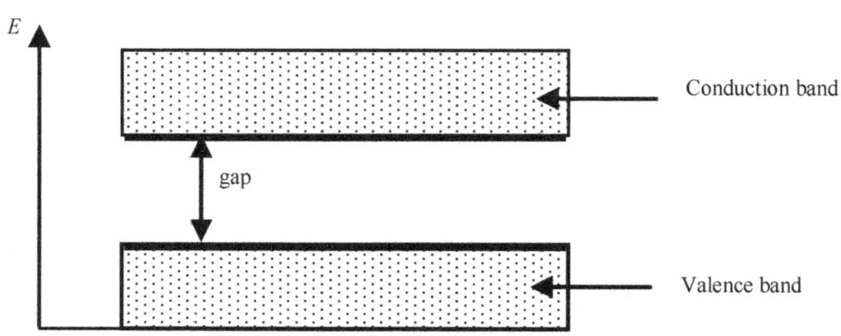

Figure 3.1. *Representation of the electron reaction on the band diagram*

3.2.2.2. *Reactions creating disorders*

We will take the example of the creation of the Schottky disorder in a binary solid, which is written as:

$$0 = nV_A + mV_B \qquad [3R.2]$$

This reaction creates a new building unit of the crystal, whose dimensions therefore increase (conservation of the ratio of sites).

3.2.2.3. *Ionization reactions*

Any structure element can, *a priori*, be ionized either by fixing an electron (or releasing an electron hole) – for example, for a vacancy of *A* which fixes an electron:

$$V_A + e' = V'_A \qquad [3R.3]$$

or by fixing an electron hole (or releasing an electron) – for example, for an interstitial atom fixing a hole:

$$A_i + h^{\circ} = A_i^{\circ} \qquad [3R.4]$$

An electron donor defect introduces a new energy level into the gap band, near to the conduction band, and ionization is the passage of the electron from the element into that band (see Figure 3.2). However, a host element is situated in the vicinity of the valence band (Figure 3.2), from which it can capture an electron.

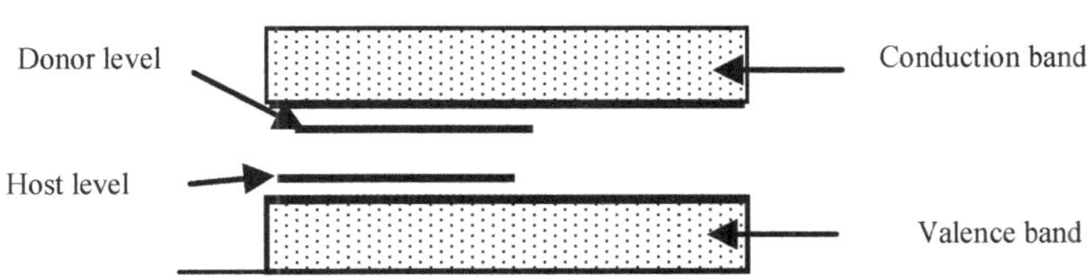

Figure 3.2. *Representation of ionization reactions in the band diagram*

3.2.2.4. *Addition reactions*

Addition reactions are reactions where structure elements combined. They are obviously very simple to write, such as in the creation of a bi-vacancy:

$$2V_A = (V_A, V_A) \qquad [3R.5]$$

Note that an ionization may be considered to be a particular addition reaction, combining a free electron or a hole with an atomic structure element.

3.2.3. *Inter-phase reactions*

Heterogeneous inter-phase reactions create or consume defects in the solid, with a different phase containing at least one element in common with the solid.

For example, we can write the reaction of creation of an atom of B in an interstitial position in the unary solid B with the gas B_2.

This creation leads to the formation of an atom of B in an interstitial position in the solid B which we will write as B_i, combined with the consumption of an interstitial vacancy V_i:

$$\frac{1}{2}B_2 + V_i = B_i \qquad [3R.6]$$

3.3. Equilibrium states between structure elements in solids

For the chemical potential (and for all the partial molar values) of a structure element, we again find the same difficulty as encountered in the case of ions. Indeed, it is impossible to maintain the quantities of elements for all types of structure elements of a solid, except for one of them because, in addition, the electrical neutrality but also the ratios between sites must be preserved in a transformation. In order to circumvent the difficulty, we use the same approach as with the ions, noting that a reaction involving structure elements must also be able to be written by using building units instead, for which the definition of a partial molar value presents no difficulty. This means that, for any intensive value (and in particular for the Gibbs energy), the expression:

$$\Delta_r X = \sum_i \nu_i \overline{X_i} \qquad [3.6]$$

retains a true meaning, even though the individual terms $\overline{X_i}$ are imaginary in value. Thus, in the context of the application of relation [3.6], we can preserve the classic expression of the chemical potential for a structure element:

$$\mu_i = g_i^o + RT \ln a_i \qquad [3.7]$$

Thus, as the expression of the affinity presents no problem, the deduction of the law of mass action to write the equilibrium conditions for a quasi-chemical reaction applies naturally.

NOTE.– Schottky adopted the reasoning of statistical thermodynamics to obtain the expression of the law of mass action in terms of fractions of sites. His idea was to use the model of the perfect solution with a mixture of structure elements within the same solid phase.

3.4. Thermodynamics of structure elements in unary solids

A solid is said to be *unary* if, in the perfect state, it contains only a single type of element which we will write as M, occupying a single type of site.

3.4.1. *Structure elements of a unary solid*

The structure elements of a unary solid may be:

– atoms or ions of M in the normal position in the lattice;

– vacant interstitial positions;

– vacancies of M, V_M, either neutral or ionized;

– atoms or ions of M in an interstitial position M_i;

– free electrons e';

– electron holes h°;

– foreign atoms or ions in substitution F_M or in insertion F_i;

– combinations of the above elements.

The first two types of structure elements are normal elements of the solid, while the others are native point defects. In general, a given solid contains several types of defects, which will be as many components in the thermodynamic sense of the term, and will form a solution with the normal elements. In practice, the problem boils down to the superposition of the equilibria of a base of the vector space. Usually, the defects are very dilute in comparison to the normal elements, so that they can be considered to be solvents with constant activity and the activities of the defects can be considered equal to their site fractions.

To construct the basis of the vector space of the balance equations, besides the electronic equilibrium between electrons and holes, we will use the equilibria of the formation of defects: the vacancies, the interstitials and the equilibria of ionization and association.

3.4.1.1. *Formation of neutral vacancies*

Neutral vacancies are formed by displacement of atoms initially in the normal position within the lattice toward normal positions on the surface of the crystal or other extended defects such as dislocations. We say that the source (and, conversely, the well) of the vacancies is an extended defect. When the surface comes into play, this results in an increase in the crystal's dimensions. If we take account of all the structure elements which are involved, the equilibrium of formation of the neutral vacancies is written as:

$$M_{M(surf)} + M_M = M_{M(surf)} + M_M + V_M + \alpha V_i \qquad [3R.7]$$

For the thermodynamic study, this formulation is simplified as:

$$0 = V_M + \alpha V_i \qquad [3R.8]$$

where α denotes the number of interstitial positions per atom of M in the crystal.

3.4.1.2. *Ionization of vacancies*

The ionization of the vacancies can be written in two different ways, depending on whether the electrons or electron holes are involved. For example, to form a vacancy once ionized negatively, we would have:

$$V_M + e' = V_M' \text{ (or } V_M = V_M' + h^{\circ}) \qquad [3R.9]$$

NOTE.– If there are no neutral vacancies, we can directly write the formation of the ionized vacancies by superposing the equilibria [3R.8] and [3R.9], in the form:

$$e' = V'_M + \alpha V_i \qquad [3R.10]$$

3.4.1.3. *Formation of neutral interstitials*

An atom which is placed in an interstitial position can only come from an atom in the normal position on the surface of the crystal (or from an extended defect), even if this displacement happens little by little. The equilibrium of formation of the interstitial is written as:

$$M + V_i = M_i \qquad [3R.11]$$

The surface vacancy that is necessarily formed is not taken into account, because in actuality, it merely expresses the fact that the displacement of the surface is accompanied by a decrease in the crystal's dimensions.

NOTE.– If we choose an atom of the volume to form the atom in an interstitial position, we would simultaneously introduce the creation of a vacancy of M.

3.4.1.4. *Ionization of the interstitials*

Similarly as for the vacancies, the ionization of the interstitials can be written in two ways, depending on whether the free electrons or the electron holes are involved. For example, to positively ionize the interstitials, we can write:

$$M_i = M_i^{\circ} + e' \text{ (or } M_i + h^{\circ} = M_i^{\circ}) \qquad [3R.12]$$

As is the case for the vacancies, the direct formation of ionized interstitials is written as follows, by superposition of the equilibria reactions [3R.11] and [3R.12] for positive interstitials:

$$M + V_i = M_i^{\circ} + e' \qquad [3R.13]$$

Of course, to each of the equilibria, we can apply the law of mass action, usually with the approximation of very dilute defects.

3.4.2. *Global equilibrium of an isolated crystal – influence of temperature*

In order to determine the global equilibrium conditions, we superimpose the partial equilibria. The equilibria that need to be taken into account depend on the nature of the defects taken into account for the solid. In general, for a given solid, the number of types of defects needing to be taken into account is relatively limited. For example, we will look at the case where six types de defects, four of which are ionized, come into play.

Thus, let us consider a solid M which, in addition to the normal elements, contains the following elements:

– vacancies of M, V_M, neutral or once ionized negatively V'_M;

– atoms of M in interstitial positions M_i, neutral or once ionized positively M_i°;

– free electrons e';

– electron holes h°.

The system of equations to be solved will involve application of the law of mass action to the equilibria [3R.1], [3R.8], [3R.9], [3R.11] and [3R.12], whose equilibrium constants are, respectively, K_e, K_V, K_1, K_n and K_2, to which we must add the expression of electrical neutrality, written as:

$$\left[h^{\circ}\right]+\left[M_i^{\circ}\right]=[V'_i]+[e'] \qquad [3.8]$$

We can see that the exact calculation quickly becomes complex. In order to simplify it, we apply Brouwer's approximation, known as the majority state approximation, to relation [3.8]: the expression of electrical neutrality. Thus, relation [3.8] will assume one of the following four extreme forms, each of which defines a temperature region:

$$\text{Region 1: } \left[h^{\circ}\right]=[e'] \quad [3.9a]$$

$$\text{Region 2: } \left[M_i^{\circ}\right]=[e'] \quad [3.9b]$$

$$\text{Region 3: } \left[M_i^{\circ}\right]=[V'_i] \quad [3.9c]$$

$$\text{Region 4: } \left[h^{\circ}\right]=[V'_i] \quad [3.9d]$$

Table 3.6 gives the expressions of the concentrations of the different defects in the different regions.

It can be remarked upon that the values of the concentrations of the neutral species do not depend on the region, which is obvious because those concentrations, which are not involved in the equation of electrical neutrality, are unaffected by the choice of a given approximation.

The limits of the regions can also be expressed as a function of the different equilibrium constants; for this purpose, we need only express two inequalities for each of the regions, so:

$$\text{Region 1: } [V'_M] \ll [e'] \text{ and } \left[M^{\circ}\right] \ll \left[h^{\circ}\right] \quad [3.10a]$$

$$\text{Region 2: } [V'_M] \ll [e'] \text{ and } \left[M^{\circ}\right] \gg \left[h^{\circ}\right] \quad [3.10b]$$

$$\text{Region 3: } [V'_M] \gg [e'] \text{ and } \left[M^{\circ}\right] \gg \left[h^{\circ}\right] \quad [3.10c]$$

$$\text{Region 4: } [V'_M] \gg [e'] \text{ and } \left[M^{\circ}\right] \ll \left[h^{\circ}\right] \quad [3.10d]$$

For each region, if we replace relation [3.8] with one of the relations [3.10], we calculate the concentrations of the different structure elements. These results are collected in Table 3.6.

Defect	Region 1	Region 2	Region 3	Region 4
$[e']$	$\sqrt{K_e}$	$\sqrt{K_2K_n}$	$\dfrac{\sqrt{K_2K_n}}{\sqrt{K_1K_V}}$	$\dfrac{\sqrt{K_e}}{\sqrt{K_1K_V}}$
$\left[h^\circ\right]$	$\sqrt{K_e}$	$\dfrac{K_e}{\sqrt{K_2K_n}}$	$K_e\dfrac{\sqrt{K_1K_V}}{\sqrt{K_2K_n}}$	$\sqrt{K_eK_1K_V}$
$[V'_i]$	$K_1K_V\sqrt{K_e}$	$K_1K_V\sqrt{K_2K_n}$	$\sqrt{K_1K_2K_nK_e}$	$\sqrt{K_1K_nK_e}$
$\left[M_i^\circ\right]$	$\dfrac{K_2K_n}{\sqrt{K_e}}$	$\sqrt{K_2K_n}$	$\sqrt{K_1K_2K_nK_e}$	$K_2K_n\sqrt{\dfrac{K_1K_V}{K_e}}$
$[V_i]$	K_V	K_V	K_V	K_V
$\left[M_i\right]$	K_n	K_n	K_n	K_n

Table 3.1. *Equilibria of a solid with six defects, four of which are charged*

If, for each region, we substitute back the expressions of the concentrations drawn from Table 3.6, we can deduce the conditions for the equilibrium constants:

$$\text{Region 1: } K_1K_V << 1 \text{ and } K_2K_n << K_e \qquad [3.11a]$$

$$\text{Region 2: } K_1K_V << 1 \text{ and } K_2K_n >> K_e \qquad [3.11b]$$

$$\text{Region 3: } K_1K_V >> 1 \text{ and } K_2K_n >> K_e \qquad [3.11c]$$

$$\text{Region 4: } K_1K_V >> 1 \text{ and } K_2K_n << K_e \qquad [3.11d]$$

It is obvious that, for a given solid, not all the conditions may be possible in its domain of stability, which would reduce the number of regions to be taken into account.

In order to express the different concentrations as a function of the temperature, we merely need to apply the van't Hoff law expressing the equilibrium constant with the temperature through the corresponding standard enthalpies.

Similarly, we could also study the equilibria of the crystal with its vapor by adding an equilibrium of phase change between the structure elements of the solid and the vapor, such as, for instance:

$$M_M = V_M + 1/2M_2(\text{gas}) \qquad [3R.14]$$

For a solid exhibiting neutral vacancies and a diatomic vapor, it is clear that the solid–vapor equilibrium chosen must be written with structure elements that are supposed to be present in the solid.

3.5. Thermodynamics of structure elements in stoichiometric binary solids

A solid is said to be binary if it ideally contains at least two structure elements occupied by atoms or ions. We sometimes encounter:

– two different chemical elements on sites of the same nature;

– two different chemical elements on two different types of sites;

– a single chemical element on two different types of sites.

The second case is, by far, the most important. It pertains, for example, to metal oxides, sulfides, halogenides, etc. Our study will pertain to this type of binary.

Our binary solid contains at least two elements, written as A and B, and at least three types of normal sites: sites of A, sites of B and at least one type of interstitial sites. In such binary compounds, we can find the following atomic defects, charged or non-charged:

– vacancies of A;

– vacancies of B;

– atoms of A in an interstitial position;

– atoms of B in an interstitial position;

– atoms of A on sites of B;

– atoms of B on sites of A.

Experience leads us to classify these solids into two families: stoichiometric solids and non-stoichiometric solids.

A solid is said to be *stoichiometric* if the ratio of the quantities of the two elements A and B remains constant and equal to its theoretical value in the perfect crystal. The ratio between the number of sites of A and the number of sites of B also remains constant. These two constants mean that we must have at least two types of defects occurring simultaneously. This set of two defects present simultaneously is known as a *disorder*. We can see from the list of defects set out above that there are six classes of disorders with two defects, which are classified into two groups: symmetrical disorders and anti-symmetrical disorders. In practice, we find only four types of disorders with two defects.

3.5.1. *Symmetrical disorders in stoichiometric binary solids*

A *disorder is symmetrical* if one of the two defects which make it up involves the lattice of the element A and the other the lattice of the element B. In practice, we find Schottky disorder and the anti-structure disorder.

Schottky disorder is the simultaneous presence of vacancies of each of the two chemical species, with these vacancies being in stoichiometric proportions to maintain the ratio between the numbers of sites.

This disorder is formed under the influence of the temperature, by transformation of the normal sites of A and B within the solid at the normal sites of A and B on the surface, which results in an increase in the dimensions of the crystal. In the case of a compound with the theoretical formula B_mA_n, the reaction of formation of the disorder can be written as:

$$0 = n\mathrm{V_A} + m\mathrm{V_B} + \alpha\mathrm{V_i} \qquad [3R.15]$$

The equilibrium constant K_s depends on the temperature, through the van't Hoff law, and the standard enthalpy of the reaction $\Delta_S H^0$.

The *anti-structure disorder* is the simultaneous presence of two types of exchanged atoms A_B and B_A. As these exchanges do not alter the stoichiometry, the equilibrium of formation of the disorder is written as follows, regardless of the values of the numbers *m* and *n* in the theoretical formula:

$$\mathrm{A_A} + \mathrm{B_B} = \mathrm{A_B} + \mathrm{B_A} \quad [3R.16]$$

The well and the source of disorder are on the actual sites. The corresponding equilibrium constant K_A also depends on the temperature on the basis of the standard enthalpy $\Delta_A H^0$.

The anti-structure disorder is encountered particularly if the two elements A and B have similar properties (comparable volumes, close electronegativity values, etc.), as happens in certain inter-metallic compounds. This is the disorder that is created during the transformation from order to long distance disorder in alloys (see Chapter 2).

3.5.2. *Asymmetrical disorders in stoichiometric binary solids*

In an *asymmetrical disorder*, the two defects that make it up pertain to the same sub-lattice of A or of B. In practice, we find only two families of anti-symmetrical disorders: Frenkel disorder and so-called "AS" disorder.

Frenkel disorder is the simultaneous presence of vacancies and interstitials of the same element. Thus, we have two possible Frenkel disorders for a binary system: Frenkel disorder on the element A and on the element B. Frenkel disorder is essentially found if there is a significant difference between the dimensions of the atoms of A and those of the atoms of B. It is for the smallest element (so as to accommodate the small atom in an interstitial position in the cavities left by the sub-lattice of the largest) that the Frenkel disorder is most likely to occur, as it is found in silver halides, which have Frenkel disorder in the silver.

Let us choose Frenkel disorder on A. Regardless of the values of the numbers *m* and *n*, the formation reaction is written as:

$$A_A + V_i = A_i + V_A \qquad [3R.17]$$

The disorder does not alter the stoichiometry of sites, so the expression of the law of mass action does not involve the coefficients *m* and *n*. The equilibrium constant $K_{F(A)}$ depends on the temperature by the standard enthalpy of formation $\Delta_{F(A)}H^0$.

The *AS disorder* is characterized by the simultaneous presence of vacancies and exchanged atoms of the same nature. Thus, we will have two possibilities of AS disorder: either on the sublattice of A with the defects V_A and A_B, or on the sublattice of B with the defects V_B and B_A. The equilibrium of formation of disorder is written as:

$$A_A = A_B + \left(1 + \frac{n}{m}\right) V_A \qquad [3R.18]$$

The creation of the disorder leads to the formation of a new building unit of the solid B_mA_n.

The possibility of the existence of atom exchange leads to necessary conditions similar to those encountered for anti-structure disorder (dimensions and electronegativities very close to those of the atoms of A and of B). This type of disorder is found essentially for inter-metallic compounds. An example is the AS disorder on the nickel in NiAl.

NOTE.– There is another disorder with two defects: the disorder of cationic distribution, which is found for binary compounds possessing two types of cationic sites. Such is the case, for example, in the spinel structure of Fe_3O_4, which has tetrahedral cationic sites and octahedral cationic sites, between which the Fe^{3+} and Fe^{2+} ions are distributed. This disorder is dealt with in the same way as the anti-structure disorder. In fact, this type of compound is closer to a ternary compound than to a binary compound, with the Fe^{3+} and Fe^{2+} ions being considered to be different components.

3.6. Thermodynamics of structure elements in non-stoichiometric binary solids

In numerous solids there are differences, sometimes very slight (of around 10^{-4}), in the ratio between the quantities of the two elements, but the ratio of the quantities of sites is still the same as in the ideal solid. These solids are said to be *non-stoichiometric*. The presence of electrons or electron holes in such ionic solids lends them the properties of semiconductors.

3.6.1. *Deviations from stoichiometry and point defects*

For a compound with the theoretical formula B_mA_n, the result of the chemical analysis shows deviations from theoretical stoichiometry m/n. The real formula can then be written in one of the following four forms, which we write in the case of an excess of B in relation to the theoretical value:

$$B_{m+\delta_B}A_n \text{ or } B_{m(1+\delta'_B)}A_n \text{ or } B_mA_{n\text{-}\delta_A} \text{ or } B_{m+\delta_B}A_n$$

This deviation from stoichiometry can also be expressed as the difference between the real ratio and the theoretical ratio:

$$\Delta = \frac{m+\delta_B}{n} - \frac{m}{n} = \frac{\delta_B}{n} = \frac{m}{n}\delta'_B \qquad [3.12]$$

To express the values Δ, δ_B or δ'_B in terms of concentrations of defects, we use the following relations:

The result of the raw chemical analysis which gives us:

$$\frac{[B]_{total}}{[A]_{total}} = \frac{m+\delta_B}{n} \qquad [3.13]$$

Generally speaking, the solid simultaneously contains vacancies of A, of B, interstitial atoms of A and B and the atoms of A and B exchanged. As all these species may be ionized to a greater or lesser degree. We use the notation $[V_B]$ to denote the total concentration of vacancies of B, ionized or otherwise. We use the same formulation for the concentrations of the other species, then the relation of conservation of sites is written as:

$$\frac{[B_B]+[A_B]+[V_B]}{[A_A]+[B_A]+[V_A]}=\frac{m}{n} \qquad [3.14]$$

The balance on the species A, which is written as:

$$[A]_{total}=[A_A]+[A_B]+[A_i] \qquad [3.15]$$

The balance on the species B which is:

$$[B]_{total}=[B_B]+[B_A]+[B_i] \qquad [3.16]$$

By combining equations [3.13], [3.14], [3.15] and [3.16], we obtain:

$$\delta_B=\frac{n\{[B_A]+[B_i]-[A_B]-[V_B]\}+m\{[A_B]+[A_i]-[B_A]-[V_A]\}}{[A_{total}]} \qquad [3.17]$$

This expression can also be written in the form:

$$\frac{\delta_B}{n}=\frac{\{[B_A]+[B_i]-[A_B]-[V_B]\}}{[A_{total}]}-\frac{m}{n}\frac{\{[A_B]+[A_i]-[B_A]-[V_A]\}}{[A_{total}]} \qquad [3.18]$$

However, if the defects are very dilute, we can accept the approximation:

$$\frac{[B]_{total}}{[A]_{total}}=\frac{m}{n} \qquad [3.19a]$$

and therefore $[A]_{total}=\frac{n}{m}[B]_{total}$ [3.19b]

Thus, equation [3.18] becomes:

$$\frac{\delta_B}{n} = \frac{m}{n}\left\{\frac{[B_A]+[B_i]-[A_B]-[V_B]}{[B_{total}]} - \frac{[A_B]+[A_i]-[B_A]-[V_A]}{[A_{total}]}\right\} \qquad [3.20]$$

This expression shows that Δ (or δ_B) is the difference between two terms:

– one accounts for the excess species B;

– the other accounts for the excess species A.

If the first term is predominant, then we would have an overall excess of B. If, however, the second term is greater, we would have an overall excess of A, in which case the value of δ_B would be negative.

3.6.2. *The predominant defect method – the Wagner classification*

Considering that in ionic compounds a defect of exchanged ions is highly improbable, and that very often, a disorder constitutes a defect of matter, and an electron defect is present in much larger amounts that other disorders, Wagner classified ionic compounds into four classes, with each class being characterized by a *predominant defect.* This approximation, and the fact that $[A_B]=[B_A]=0$, simplify relation [3.20]. In every case, the electronic part of the disorder (electrons or electron holes) can be free or trapped on other structure elements, normal elements or atomic defects.

3.6.2.1. *Compounds with predominant anionic vacancies*

Such compounds are characterized by the following concentrations:

$$[V_A]\neq 0 \quad [V_B]=[B_i]=[A_i]=0 \qquad [3.21]$$

The deviation from stoichiometry is then simply expressed as:

$$\Delta = \frac{m}{n}\frac{[V_A]}{[A]_{total}} \cong \frac{m}{n}\frac{[V_A]}{[A_A]} \qquad [3.22]$$

Electrical neutrality is assured by an electronic part made up of free or trapped electrons, which lends this group the properties of a semiconductor *n*. Cerium oxide CeO_2 is an example of a substance belonging to this family.

3.6.2.2. *Compounds with predominant interstitial cations*

Such compounds are characterized by the following concentrations:

$$[B_i] \neq 0 \quad [V_B] = [V_A] = [A_i] = 0 \qquad [3.23]$$

The deviation from stoichiometry is then simply expressed as:

$$\Delta = \frac{m}{n}\frac{[B_i]}{[B]_{total}} \cong \frac{m}{n}\frac{[B_i]}{[B_B]} \qquad [3.24]$$

Electrical neutrality is assured by an electronic part composed of free or trapped electrons, which lends this group the properties of a semiconductor *n*. We can cite barium oxide BaO as an example of a substance that belongs to this family.

3.6.2.3. *Compounds with predominant cationic vacancies*

Such compounds are characterized by the following concentrations:

$$[V_B] \neq 0 \quad [V_A] = [B_i] = [A_i] = 0 \qquad [3.25]$$

The deviation from stoichiometry is then simply expressed as:

$$\Delta = \frac{m}{n}\frac{[V_B]}{[A]_{total}} \cong \frac{m}{n}\frac{[V_B]}{[B_B]} \qquad [3.26]$$

Electrical neutrality is ensured by an electronic part made up of free or trapped electron holes, giving this group semiconductive properties *p*. We can cite iron (II) oxide FeO as belonging to this family.

3.6.2.4. *Compounds with predominant interstitial anions*

Such compounds are characterized by the following concentrations:

$$[A_i] \neq 0 \quad [V_B] = [B_i] = [V_A] = 0 \qquad [3.27]$$

The deviation from stoichiometry is then simply expressed as:

$$\Delta = \frac{m}{n}\frac{[A_i]}{[A]_{total}} \cong \frac{m}{n}\frac{[A_i]}{[A_A]} \qquad [3.28]$$

Electrical neutrality is assured by an electronic part made up of free or trapped electron holes, which gives this family the properties of a semiconductor *p*. For example, we can cite uranium oxide UO_2 as belonging to this family.

Ultimately, the Wagner solid is characterized by a disorder with an atomic defect and an electronic defect. Table 3.7 shows the four Wagner cases, each time specifying the sign of the electronic defect and the type of semiconductor.

Predominant defect	Sign of the charge carrier	Type of semiconductor
Anion vacancies	–	n
Interstitial cations	–	n
Cation vacancies	+	p
Interstitial anions	+	p

Table 3.2. *The four Wagner cases of the predominant defect*

3.6.3. *Equilibrium of a Wagner solid with one of its gaseous elements*

The Wagner approximation is sometimes sufficient to study the defects of a solid in a chemical equilibrium. For example, we can easily study the equilibrium of a Wagner solid with an oxidizing or reducing gas, in pressure domains which maintains the solid in question stable. Let us look, for instance, at the equilibrium of the defects in cerium oxide with the oxygen pressure in the pressure domain of stability of that oxide. We can see that, because cerium oxide is a solid with anionic vacancies, an increase in oxygen pressure brings the solid closer to stoichiometric composition. The quasi-chemical reaction with oxygen can be written in the form:

$$\mathrm{O_O} + \mathrm{Ce_{Ce}} = \mathrm{V_O^{\circ\circ}} + 2\mathrm{Ce'_{Ce}} + 1/2\mathrm{O_2} \qquad [3R.19]$$

The application of the law of mass action to that equilibrium, with the hypothesis of ideal solutions, gives us:

$$K = \left[\mathrm{V_O^{\circ\circ}}\right]\left[\mathrm{Ce'_{Ce}}\right]P_{\mathrm{O_2}}^{1/2} \qquad [3.29]$$

We can easily deduce the expression of the concentrations of defects as a function of the oxygen pressure, so:

$$2\left[\mathrm{V_O^{\circ\circ}}\right] = \left[\mathrm{Ce'_{Ce}}\right] = (2K)^{1/3}\, P_{\mathrm{O_2}}^{-1/6} \qquad [3.30]$$

The concentration of defects decreases when the oxygen pressure increases.

The reasoning would be the same for a reaction in the presence of a reducing gas – e.g. with hydrogen, an increase in the pressure of that gas will take the solid further away from stoichiometry. The reaction is written as:

$$1/2\mathrm{H_2} + \mathrm{O_O} + \mathrm{Ce_{Ce}} = \mathrm{V_O^{\circ\circ}} + 2\mathrm{Ce'_{Ce}} + \mathrm{H_2O} \qquad [3R.20]$$

The application of the law of mass action enables us to immediately formulate the variations in the concentrations of defects with the ratio of the partial pressures of hydrogen and water vapor. Thus, we can show that the concentrations of defects increase when we raise the partial pressures of hydrogen and water vapor.

3.6.4. *General equilibrium of a non-stoichiometric binary solid with one of its gaseous elements*

In the hypothetical case where the approximation of the predominant Wagner defect is no longer sufficient, the solid will contain several atomic defects, and the examination of its behavior in an equilibrium state will involve the superposition of chemical equilibria – some internal to each of the phases present, and others expressing inter-phase transfers. We can see that we soon find complex systems of the same type as those encountered with ionic equilibria in an aqueous solution. However, to formulate approximate solutions, Brouwer used the approximation of the majority defect which we already encountered in section 3.4.2.

For example, consider a compound AB exhibiting the following defects:

$$\mathrm{A_A, B_B, V_A, V_A^{'}, A_i, A_i^{\circ}, e' \text{ and } h^{\circ}}$$

We wish to study the equilibrium of that compound with the gas B_2.

The deviation from stoichiometry is linked to the concentrations of defects, in line with the use of relation [3.20], as follows:

$$\delta = \mathrm{A_i + A_i^{\circ} - V_A - V_A^{'}} \qquad [3.31]$$

We are going to determine the concentrations of the different defects, in isothermal conditions, as a function of the partial pressure of the gas B_2. In order to do so, we will look at the different equilibria.

3.6.4.1. *Internal equilibria in the solid*

Let us write the independent equilibria of formation of the defects.

– formation of a neutral vacancy of A and an interstitial atom A (Frenkel disorder):

$$0 = \mathrm{A_i} + \mathrm{V_A} \text{ with } K_F = \left[\mathrm{A_i^\circ}\right]\left[\mathrm{V_A}\right] \quad [3.32]$$

– ionization of the vacancy of A:

$$\mathrm{V_A} + \mathrm{e'} = \mathrm{V_A'} \text{ with } K_1 = \frac{\left[\mathrm{V_A'}\right]}{\left[\mathrm{V_A}\right]\left[\mathrm{e'}\right]} \quad [3.33]$$

– ionization of the interstitial atom A:

$$\mathrm{A_i} + \mathrm{h^\circ} = \mathrm{A_i^\circ} \text{ with } \mathrm{K_2} = \frac{\left[\mathrm{A_i^\circ}\right]}{\left[\mathrm{A_i}\right]\left[\mathrm{h^\circ}\right]} \quad [3.34]$$

– electronic equilibrium:

$$0 = \mathrm{e'} + \mathrm{h^\circ} \text{ with } K_i = \left[\mathrm{e'}\right]\left[\mathrm{h^\circ}\right] \quad [3.35]$$

To these equations, we have to add the condition of electrical neutrality:

$$\left[\mathrm{V_A'}\right] + \left[\mathrm{e'}\right] = \left[\mathrm{h^\circ}\right] + \left[\mathrm{A_i^\circ}\right] \quad [3.36]$$

3.6.4.2. *Transfer of atoms between the two phases*

We will write an inter-phase equilibrium which involves only the defects taken into account in the above equilibria. Let us choose the neutral vacancy of A, which enables us to write the equilibrium in the form:

$$1/2\mathrm{B_2} = \mathrm{B_B} + \mathrm{V_A} \text{ with } K_{BV} = \frac{\left[\mathrm{V_A}\right]}{\mathrm{P_{B_2}^{1/2}}} \quad [3.37]$$

We can easily show that any other transfer equilibrium is a linear combination of the above one and the internal equilibria.

3.6.4.3. *Solving the system*

We have six independent relations [3.32], [3.33], [3.34], [3.35], [3.36] and [3.37] for seven unknowns, which are:

$$[\mathrm{V_A}],\left[\mathrm{V_A'}\right],[\mathrm{A_i}],\left[\mathrm{A_i^\circ}\right],\ [\mathrm{e'}],\ \left[\mathrm{h^\circ}\right] \text{ and } P = P_{\mathrm{B_2}}$$

Thus, we can express all the unknowns as a function of one of them – e.g. the pressure P of B_2.

In fact, the system thus obtained is complex to solve, particularly when we do not have the numerical values of the equilibrium constants. We can simplify the calculation by using Brouwer's majority defect approximation, which, remember, involves considering only the two largest terms (one with each sign) in the expression of electrical neutrality [3.36]. Thus, in our example, we would have four Brouwer cases, defined by:

Case 1: $\left[\mathrm{A_i^\circ}\right] = [\mathrm{e'}]$ Case 2: $[\mathrm{e'}] = \left[\mathrm{h^\circ}\right]$

Case 3: $\left[\mathrm{V_A'}\right] = \left[\mathrm{A_i^\circ}\right]$ Case 4: $\left[\mathrm{V_A'}\right] = \left[\mathrm{h^\circ}\right]$

NOTE.– In each case, these expressions define the majority defects, which may not necessarily be the predominant defects in Wagner's sense. Indeed, the neutral defects, which do not appear in the expression of electrical neutrality, may be present in larger amounts than the majority defects and become predominant.

Table 3.8 gives the expressions of the concentrations of each of the defects. Each Brouwer case corresponds to a pressure domain, which we will write explicitly. For this purpose, let us set:

$$R_1 = \frac{1}{K_1 K_{BV}} \quad [3.38a]$$

and $$R_2 = \frac{K_2 K_F}{K_{BV}} \quad [3.38b]$$

If we replace the concentrations in the inequalities, we obtain the following conditions:

$$\text{Case 1:}\begin{cases}[\mathrm{e}'] >> [\mathrm{V}_\mathrm{A}^{'}]\\ [\mathrm{h}^\circ] << [\mathrm{A}_\mathrm{i}^\circ]\end{cases} \text{so} \begin{cases}P << R_1^2\\ P << R_2^2\end{cases}; \quad \text{Case 2:}\begin{cases}[\mathrm{e}'] >> [\mathrm{V}_\mathrm{A}^{'}]\\ [\mathrm{h}^\circ] >> [\mathrm{A}_\mathrm{i}^\circ]\end{cases} \text{so} \begin{cases}P << R_1^2\\ P >> R_2^2\end{cases}$$

$$\text{Case 3:}\begin{cases}[\mathrm{e}'] << [\mathrm{V}_\mathrm{A}^{'}]\\ [\mathrm{h}^\circ] << [\mathrm{A}_\mathrm{i}^\circ]\end{cases} \text{so} \begin{cases}P >> R_1^2\\ P << R_2^2\end{cases}; \quad \text{Case 4:}\begin{cases}[\mathrm{e}'] << [\mathrm{V}_\mathrm{A}^{'}]\\ [\mathrm{h}^\circ] >> [\mathrm{A}_\mathrm{i}^\circ]\end{cases} \text{so} \begin{cases}P >> R_1^2\\ P >> R_2^2\end{cases}$$

We can see that the four cases are not possible simultaneously for the same solid, and that we must envisage two possibilities depending on the relative values of R_1 and R_2 at the temperature in question:

$$\text{1st possibility: if } R_2 > R_1 \tag{3.39}$$

If the pressure P increases, we successively encounter cases 1, 3 and 4.

Species	Region 1	Region 2	Region 3	Region 4
$[\mathrm{e}']$	$\sqrt{\frac{K_2 K_i K_F}{K_{BV}}} P^{-1/4}$	$\sqrt{K_i}$	$\frac{1}{K_{BV}}\sqrt{\frac{K_F K_i K_2}{K_1}} P^{-1/2}$	$\sqrt{\frac{K_i}{K_1 K_{BV}}} P^{-1/4}$
$[\mathrm{h}^\circ]$	$\sqrt{\frac{K_{BV} K_i}{K_2 K_F}} P^{1/4}$	$\sqrt{K_i}$	$K_{BV}\sqrt{\frac{K_i K_1}{K_F K_2}} P^{1/2}$	$\sqrt{K_i K_1 K_{BV}}\, P^{1/4}$
$[\mathrm{V}_\mathrm{A}^{'}]$	$K_1\sqrt{K_i K_F K_2 K_{BV}}\, P^{1/4}$	$K_1 K_{BV}\sqrt{K_i}\sqrt{P}$	$\sqrt{K_i K_F K_2 K_1}$	$\sqrt{K_i K_1 K_{BV}}\, P^{1/4}$
$[\mathrm{A}_\mathrm{i}^0]$	$\sqrt{\frac{K_{BV} K_i}{K_2 K_F}} P^{1/4}$	$\frac{K_2 K_F \sqrt{K_i}}{K_{BV}} P^{-1/2}$	$\sqrt{K_i K_F K_2 K_1}$	$K_2 K_F\sqrt{\frac{K_i K_1}{K_{BV}}} P^{-1/4}$
$[\mathrm{V}_\mathrm{A}]$	$K_{BV}\sqrt{P}$	$K_{BV}\sqrt{P}$	$K_{BV}\sqrt{P}$	$K_{BV}\sqrt{P}$
$[\mathrm{A}_\mathrm{i}]$	$\frac{K_F}{K_{BV}\sqrt{P}}$	$\frac{K_F}{K_{BV}\sqrt{P}}$	$\frac{K_F}{K_{BV}\sqrt{P}}$	$\frac{K_F}{K_{BV}\sqrt{P}}$

Table 3.3. *Concentrations of defects at equilibrium of a binary with a gas*

$$\text{2nd possibility: if } R_2 < R_1 \tag{3.40}$$

Only the succession of cases 1, 2 and 4 will be possible.

Figure 3.3 illustrates the curves giving the concentrations of the different species as a function of the pressure of B_2 in the case where $R_2 < R_1$. The diagram obtained is known as a Kröger–Vink diagram. The diagram is plotted on the basis of the approximate expressions given in Table 3.8. In theory, the true curves are not straight lines, so strictly speaking, the lines drawn here are merely asymptotes to the real curves. The breaks which appear in the diagram are solely due to Brouwer's approximations.

Note that the zone furthest to the left in Figure 3.3 corresponds to the case of the Wagner solid with an interstitial cation A and free electrons, while the zone furthest to the right corresponds to the case of a Wagner solid with a cationic vacancy and free electron holes. Thus, we can show the possibility, with the same solid, of moving from a semiconductor n to a semiconductor p merely under the influence of the gas pressure.

NOTE.– For a given system with a given solid and gas, the pressure domain covered depends on the stabilities of the system and is, in fact, limited by equilibria of phase change such as the condensation of gas B_2, the stability of the compound AB, etc. Hence, in real cases, it is extremely rare to obtain three zones for the same system.

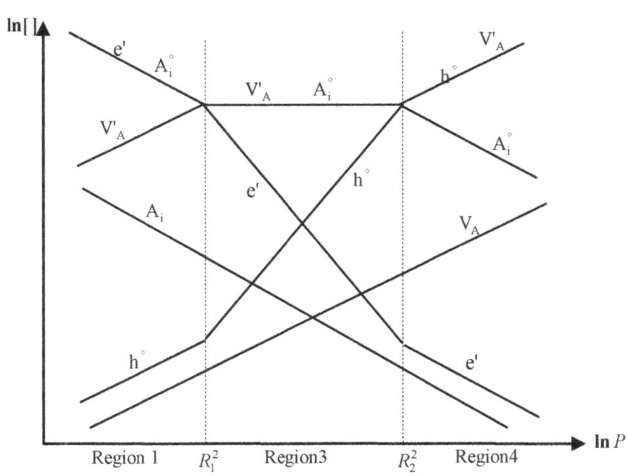

Figure 3.3. *Representation of a Kröger–Vink diagram*

3.7. Representation of complex solids – example of metal oxy-hydroxides

We can easily imagine that increasing the complexity in the description of a solid in its structure elements when we switch from binary solids to ternary, quaternary (etc.) solids. Certain simplification methods are used to correctly model the behavior of solids more complex than binaries. These degeneration methods enable us to decrease the number of structure elements taken into account. We will cite two fairly common examples.

3.7.1. *The pseudo-binary approximation*

In a solid, there may be different kinds of bonds between the atoms, meaning that we can group together several atoms bound together in a single structure element. Such is the case with ionic compounds containing complex ions: the atoms which make them up are linked by covalent bonds, and each complex ion will behave like a unique element on a unique site.

Such is the case, for example, with metal carbonates in which we can define the metal cations, on the one hand, and the carbonate anions on the other hand, as normal structure elements. We will not distinguish the behavior of the individual oxygen atoms. Thus, these compounds can be considered to be binaries, so this approximation is referred to as "pseudo-binary". These compounds may, of course, contain defects, such as an oxygen ion on an anionic site which is normally occupied by a carbonate ion.

3.7.2. *The predominant-defect generalization*

As Wagner did with binary compounds, in a more complex solid, it is possible to take account only of the defects deemed to be predominant and describe the solid in view only of those defects.

Let us take the example of metal oxy-hydroxides such as boehmite (AlOOH). We can describe the ideal solid using the following structure elements:

Al_{Al}, OH_{OH} and O_O

True boehmite exhibits defects in relation to the sites of the hydroxyl groups. The defect of an oxygen ion in substitution of OH (which must not be confused with an oxygen ion in the oxygen position) is taken to be the predominant defect. So that electrical neutrality is respected, thus results in the presence of vacancies of OH^- ions, such that we have:

$$\left[O_{OH}^{'}\right]=\left[V_{OH}^{\circ}\right] \qquad [3.41]$$

Because of the conservation of the sites of the OH ions, we need the relation:

$$\left[O_{OH}^{'}\right]+\left[V_{OH}^{\circ}\right]+\left[OH_{OH}\right]=1 \qquad [3.42]$$

Hence, for the description of the properties of boehmite linked to the loss of water, we do not need to worry about the aluminum ions and oxygen ions placed in their normal positions. The departure of the water is expressed by the reaction:

$$2OH_{OH}=H_2O+V_{OH}^{\circ}+O_{OH}^{-} \qquad [3R.21]$$

By studying this equilibrium, we can find the concentrations of the different structure elements in question, as a function of the water vapor pressure.

The concept of a structure element and that of non-stoichiometry are most important in understanding phenomena involving solids, such as electrical conductivity, the mechanisms of heterogeneous reactions and certain optical and magnetic properties.

3.8. Determination of the equilibrium constants of the reactions involving structure elements

All quasi-chemical reactions involving structure elements, and in particular, point defects (which we saw above), obviously involve equilibrium constants that need to be determined. In most cases, these constants are calculated on the basis of data gleaned by statistical thermodynamics.

3.8.1. *Recap on calculating the equilibrium constants using statistical thermodynamics*

The reactions which we have encountered, with the exception of the inter-phase reactions, take place in the solid phase, so all the components in the reaction are components of the same solid solution.

For such a solution, the variation in volume due to the reaction is negligible, and therefore we will have:

$$\Delta_r G(T) = \Delta_r F(T) + P\Delta_r V \cong \Delta_r F(T) \qquad [3.43]$$

We can express the Helmholtz energy using statistical thermodynamics, remembering that the molecules of solid are considered to be discernible molecules. If we have a population of N_i molecules of the species i which each have a molecular partition function z_i, then the Helmholtz energy of that species i is written as:

$$N_i \;\; F_i(T) - F_i(0) = -N_i \mathrm{k_B} T \ln z_i \qquad [3.44]$$

In this expression, the solution is taken to be perfect, because no account is taken of an enthalpy of mixing.

In standard conditions, at temperature T, the Helmholtz energy of n moles of the component i becomes:

$$f_i^0(T) = f_i^0(0) - n_i \mathrm{R} T \ln z_i \qquad [3.45]$$

$f_i^0(0)$ is the standard molar Helmholtz energy of the pure component i at the temperature of 0 K. At this temperature, the entropy of all the species is zero, and therefore the Helmholtz energy is identical to the internal energy which, in view of approximation [3.43], is identical to the enthalpy at 0 K, so:

$$f_i^0(0) \cong u_i^0(0) \qquad [3.46]$$

For a reaction in which the stoichiometric numbers are a_i, we write:

$$\Delta_r f^0(T) = \sum_i a_i f_i^0(0) = \mathrm{R}T \sum_i a_i \ln z_i \qquad [3.47]$$

By definition, and in view of relation [3.43], the equilibrium constant is:

$$-\mathrm{R}T \ln K_x = \Delta_r g^0(T) = \Delta_r f^0(T) \qquad [3.48]$$

Using relations [3.46] and [3.47], we find:

$$-\mathrm{R}T \ln K_x = \Delta_r \left[u_i^0(0) \right] - \mathrm{R}T \ln \prod_i (z_i)^{a_i} \qquad [3.49]$$

Thus, for the equilibrium constant relative to the molar fractions:

$$K_x = \prod_i (z_i)^{a_i} \exp\left(-\frac{\Delta_r u_0}{\mathrm{R}T} \right) \qquad [3.50]$$

In the case of solids, the only terms included in the partition function are:

– the coefficient of multiplicity of the electron state g_{e_i}, which usually takes the value of 1 but, in the presence of charged species which have one free electron, takes the value of 2 (two spin states);

– the vibration term which, in most cases, is simply written in the form:

$$z_v = \frac{\mathrm{k_B} T}{\mathrm{h}\nu_i} \qquad [3.51]$$

The frequency ν is the vibrational frequency of the lattice in the Einsteinian approximation.

If we substitute these values back into relation [3.50], the equilibrium constant (relative to the molar fractions) becomes:

$$K_x = \prod_i \left(g_{e_i} \frac{\mathrm{k_B} T}{\mathrm{h}\nu_i} \right)^{a_i} \exp\left(-\frac{\Delta_r u_0}{\mathrm{R}T} \right) \qquad [3.52]$$

We can see that this relation is the product of two terms: the pre-exponential term and the variation in internal energy due to residual vibrations at the temperature of 0 K.

NOTE.– The equilibrium constant expressed by relation [3.52] is that which pertains to the activity levels. If we use the concentrations, we can easily show that, for very dilute solutions, which is generally the case for point defects:

– the activity coefficient relative to the concentrations (molar-solution reference) is practically equal to the activity coefficient in the pure-substance reference;

– if we let v_0^0 denote the molar volume of the solvent the solid here), the equilibrium constant relative to the concentrations is expressed, on the basis of the constant K_x, by the relation:

$$K_c = K_x \left(v_0^0\right)^{\sum_i \nu_i} \qquad [3.53]$$

3.8.2. *Examination of the pre-exponential term in the quasi-chemical equilibrium constants*

To begin with, we can accept the hypothesis that a point defect does not affect the lattice's fundamental vibration frequency, and therefore the term ν in relation [3.51] does not depend on the species involved in the quasi-chemical reaction. For ionic compounds, we sometimes choose one anionic vibration frequency and one cationic frequency. Certain authors, such as Mott [MOT 38], opt instead for a half frequency for the defect. We will now examine a few examples, with the vibration frequency being kept constant and unique.

3.8.2.1. *Creation of a vacancy*

The reaction is written as:

$$0 \rightarrow \mathrm{V} \qquad [3R.22]$$

Remembering that the “0” on the left-hand side of the above reaction actually represents the ideal solid, the pre-exponential term would be:

$$K_{\mathrm{V}}^{0}=\frac{\mathrm{k_B}T}{\mathrm{h}\nu_i}\left(\frac{\mathrm{k_B}T}{\mathrm{h}\nu_i}\right)^{-1}=1 \qquad [3.54]$$

3.8.2.2. *Creation of an interstitial atom*

The reaction is written as:

$$\underset{(\text{perfect critical})}{\mathrm{M_M}+\mathrm{V_i}}\Leftrightarrow \mathrm{M_i} \qquad [3R.23]$$

The pre-exponential term would be:

$$K_{\mathrm{V}}^{0}=\frac{\mathrm{k_B}T}{\mathrm{h}\nu_i}\left(\frac{\mathrm{k_B}T}{\mathrm{h}\nu_i}\right)^{-1}=1 \qquad [3.55]$$

3.8.2.3. *Association of two vacancies*

The reaction is written as:

$$2\mathrm{V}\rightarrow(\mathrm{V})_2 \qquad [3R.24]$$

Hence, the pre-exponential term would be:

$$K_{\mathrm{V}}^{0}=Z\left(\frac{\mathrm{k_B}T}{\mathrm{h}\nu_i}\right)\left(\frac{\mathrm{k_B}T}{\mathrm{h}\nu_i}\right)^{-2}=Z\left(\frac{\mathrm{k_B}T}{\mathrm{h}\nu_i}\right)^{-1} \qquad [3.56]$$

Z is the number of near neighbors of a vacancy, because a bivacancy has Z equivalent positions, which gives us the multiplicity coefficient Z for the vibrational partition function – e.g. 6 for a cubic centered lattice.

3.8.2.4. *Ionization of a vacancy*

The reaction is written as:

$$\mathrm{V_i}+\mathrm{e}'\rightarrow \mathrm{V}' \qquad [3R.25]$$

Thus, the pre-exponential term would be:

$$K_{\mathrm{V}}^{0}=1\left(\frac{\mathrm{k_B}T}{\mathrm{h}\nu_i}\right)\left(\frac{\mathrm{k_B}T}{\mathrm{h}\nu_i}\right)=\left(\frac{\mathrm{k_B}T}{\mathrm{h}\nu_i}\right)^{-1} \qquad [3.57]$$

The coefficient 1 represents the assumption that the ionized vacancy has one free electron.

3.8.2.5. *Creation of a Schottky disorder*

The reaction is written as:

$$0 \rightarrow \mathrm{V_A+V_B} \qquad [3R.26]$$

Thus, the pre-exponential term would be:

$$K_{\mathrm{V}}^{0}=\left(\frac{\mathrm{k_B}T}{\mathrm{h}\nu_i}\right)\left(\frac{\mathrm{k_B}T}{\mathrm{h}\nu_i}\right)^{-1}=\left(\frac{\mathrm{k_B}T}{\mathrm{h}\nu_i}\right) \qquad [3.58]$$

3.8.2.6. *Creation of a Frenkel disorder*

The reaction is written as:

$$\mathrm{M+V_i \rightarrow M_i+V_M} \qquad [3R.27]$$

Hence, the pre-exponential term would be:

$$K_{\mathrm{V}}^{0}=\left(\frac{\mathrm{k_B}T}{\mathrm{h}\nu_i}\right)^{2}\left(\frac{\mathrm{k_B}T}{\mathrm{h}\nu_i}\right)^{-2}=1 \qquad [3.59]$$

3.8.2.7. *Inter-phase exchange reaction*

Let us examine the equilibrium between an atom B of the lattice and the gas B_2. The reaction is written as:

$$\mathrm{B_B \rightarrow V_B+1/2[B_2]} \qquad [3R.28]$$

In the case of a gas, it is wise to take account of the translational partition function expressed in relation to the molar volume and the rotational partition function (with the moment of inertia *I*). The pre-exponential term becomes:

$$K_{\mathrm{V}}^{0} = \left(\frac{\mathrm{k_B}T}{\mathrm{h}\nu_i}\right)\left(\frac{\mathrm{k_B}T}{\mathrm{h}\nu_i}\right)^{-1}\left[v_B^0\left(\frac{2\pi m\mathrm{k_B}T}{\mathrm{h}^2}\right)^{3/2}\frac{4\pi^2 I\mathrm{k_B}T}{\mathrm{h}^2}\right]^{1/2}$$
$$= \left[v_B^0\left(\frac{2\pi m\mathrm{k_B}T}{\mathrm{h}^2}\right)^{3/2}\frac{4\pi^2 I\mathrm{k_B}T}{\mathrm{h}^2}\right]^{1/2} \qquad [3.60]$$

The term m is the mass of the molecule B_2 and v_B^0 is its molar volume.

3.8.2.8. *Case of electron equilibrium*

The electron equilibrium reaction is written as:

$$0 \rightarrow \mathrm{e'} + \mathrm{h}^{\circ} \qquad [3R.29]$$

Fundamentally, electrons and holes behave like monatomic gases in the volume of the solid. That being the case, the partition function includes only the molar translation term, which immediately gives us the pre-exponential term:

$$K_{\mathrm{e}}^{0} = 4\left(\frac{2\pi m\mathrm{k_B}T}{\mathrm{h}^2}\right)^{3} \qquad [3.61]$$

The factor 4 comes from the two electron statistical weights for the two levels of spin, and m is the real mass of the electron.

3.8.3. *Determination of the internal energy of transformation of quasi-chemical reactions*

The energy term included in the exponential of relation [3.52] is the transformation energy envisaged at 0 K. This energy can be deduced from experience, but the energy values relative to point defects are often extremely tricky or difficult to implement. It is in this context that the importance of prior calculation of those energies becomes apparent. Of the various energies of defect creation, we are going to envisage two cases: the energy of creation of a defect, however, and the energy of creation of a disorder which, by definition, includes multiple defects. We will begin by looking at this second case.

3.8.3.1. *Energy of creation of a disorder*

The energy of creation of a disorder is determined on the basis of a Born–Haber cycle. We will look at the example of formation of the doubly-ionized Schottky disorder. We envisage the following Born–Haber cycle:

$\mathrm{A_A \rightarrow A^+(gas) + V_a^{'}}$ $\left(H_{\mathrm{A^+}}\right)$: enthalpy of creation of the defect V'$_\mathrm{A}$

$\mathrm{B_B \rightarrow B^-(gas) + V_B^{\circ}}$ $\left(H_{\mathrm{B^-}}\right)$: enthalpy of creation of the defect V°$_\mathrm{B}$

$\mathrm{A^+(gas) + e^- \rightarrow A(gas)}$ $\left(\text{-}I_\mathrm{A}\right)$: ionization energy of A)

$\mathrm{B^-(gas) \rightarrow B(gas) + e^-}$ $\left(\text{-}E_B\right)$: electronic affinity of B

$\mathrm{A(gas) + B(gas) \rightarrow A_A + B_B}$ $\left(H_{\mathrm{AB}}\right)$: enthalpy of formation of the solid AB from its gaseous elements

$0 \rightarrow \mathrm{V_A^{'} + V_B^{\circ}}$ $\left(H'_S\right)$: enthalpy of formation of the doubly-ionized Schottky disorder

From this, we can deduce the relation:

$$H'_S = -I_\mathrm{A} - E_\mathrm{B} + H_{\mathrm{A^+}} + H_{\mathrm{B^-}} + H_{\mathrm{AB}} \qquad [3.62]$$

The terms in this sum are known, with the exception of the two enthalpies $\left(H_{\mathrm{A^+}}\right)$ and $\left(H_{\mathrm{B^-}}\right)$ of defect creation. Thus, we are led to the determination of the enthalpies of defect creation.

NOTE.– In order to use expression [3.52], we must take the enthalpies of the different reactions at 0 K.

3.8.3.2. *Determination of the defect creation energies*

Let us begin by saying that calculating the energy of creation of a defect is a complex operation, normally carried out by specialists in automated numerical computation. For this reason, we will content ourselves with outlining the methods used without going into the technical details of the computation.

It is clear that if we wish to calculate, say, the energy of creation of a vacancy, we need to be able to calculate a lattice energy of a solid with and without the defect. The difference between these two terms would give us the energy of formation of the defect. We now begin to see the first difficulty: the result we are seeking is the difference between two significant numbers, so obviously, we need a precise measure of each of them.

If, indeed, we proceed in this way, contenting ourselves with removing an atom or an ion to create the vacancy, the result is disastrous. Indeed, this type of crude calculation gives us an energy of formation of around 8eV per molecule, while experience tells us that the real value is 2eV per molecule.

NOTE.– The unit used by specialists in calculating these energies is the electron-volt (eV) per molecule, which equates to 96.48 kJ/mole.

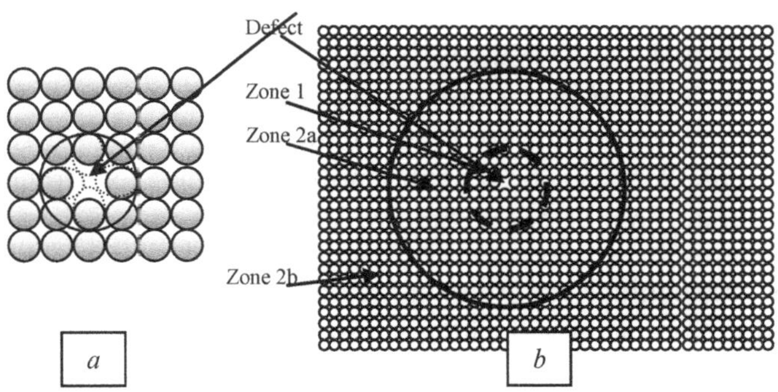

Figure 3.4. *2D representation of the crystal, showing a) the relaxation of the ions around the vacancy; b) Mott and Littleton's calculation zones*

The difference observed here arises from two phenomena. When the atom is removed, the neighboring atoms move and deform. These phenomena are known as relaxation and polarization (Figure 3.4(a)) of the atoms or ions, and the energy corresponding to these phenomena is, indeed, around 6eV.

Thus, we can see that in order to calculate the energy of creation of the defect, it is clearly necessary to have a potential function but also to be capable of taking account of the relaxations and polarization of the atoms or ions placed around the defect.

The potential functions we have were constructed by combination of the following interactions between two atoms or two ions:

– electrostatic forces between ions, often assimilated to points. These are forces that are exerted over a long distance (in $1/r^2$, and the energy is in $1/r$);

– the Lennard–Jones forces of attraction, exerted between atoms (charged or otherwise), at a short distance (energy in $1/r^6$);

– the forces of repulsion resulting from the interpenetration of the electron clouds of neighboring atoms, exerted at a very short distance (energy in $1/r^{12}$ or in exponentials of r).

The various methods used, in fact, belong to two families: *supercell* methods and *fixed cluster* methods, which are also called *force equilibrium methods*. Both types of methods have their advantages and disadvantages, with the supercell method possibly being better adapted in the case of solutions with a high concentration of defects, whereas the fixed-cluster method is better suited for the case of dilute defects. The second family of methods is perhaps a little more accurate, so it is that method which we will describe.

The force equilibrium group of methods are all drawn from a famous publication from Mott and Littleton [MOT 38]. This family is therefore known as the ML method, which, over time, has undergone certain modifications which tend to increase the precision of the calculations and yield the following methods: Point Polarizable Ion (PPI) model, Modified Point Polarizable Ion (MPI) model and Extended Polarizable Point Ion (EPPI) model. For example, Mott and Littleton only took account of the electrical forces and the forces of repulsion. The other forces have been added in order to improve the results.

The ML method is based on the two-region strategy. We choose a central point, or origin, which is situated at the center of the defect. If the defect is composed of multiple entities, the central point is identical to the center of those entities. Around the defect, the space is divided into two spherical regions (Figure 3.4(b)): the inner sphere is called region 1, and the space between the two spheres constitutes region 2a. The rest of the space, of supposedly infinite dimensions, is called region 2b. The dimensions of the spheres are defined by their radius and the number of atoms (or ions) they contain. Of course, this number must be evaluated. If it is too high, the computation time becomes prohibitive, but too low a number will very quickly afflict the precision of the calculations. Catlow estimates that region 1 should typically include 100 atoms (or ions).

In the earliest versions of the method, which emerged when we did not yet have the resources currently at our disposal, only the first region was treated as being composed of discrete entities (6 atoms to begin with); regions 2a and 2b were treated as dielectric elastic continua.

Region 1 will be considered to be greatly disturbed by the defect, and the ions it contains can be displaced to new positions that eliminate the forces to which they are subject. However, regions 2a and 2b will be considered to be only slightly disturbed, and their displacements can be determined approximately.

To begin with, we will combine regions 2a and 2b into a unified, unique region 2. Let x be the Cartesian coordinates of the atoms and ξ the displacements in Cartesian coordinates. Then we can express the total energy of the system as the sum of the energies of each of the regions 1 and 2 and of the energy of interaction between them, which we will formulate as follows:

$$U_{tot}(x,\xi) = U_{11}(x) + U_{12}(x,\xi) + U_{22}(\xi) \quad [3.63]$$

If the forces acting on region 2 are slight, we can suppose that the region supports the hypothesis of harmonicity, and therefore the energy of region 2 will be written as:

$$U_{22}(x,\xi) = \frac{1}{2}\xi \mathrm{H}_{22}\xi \quad [3.64]$$

In this expression, H_{22} is the Hessian matrix for region 2 – i.e. the matrix of the second partial derivatives in relation to the displacements. If we now write that the displacements in region 2 are at equilibrium, then we need to satisfy the following condition:

$$\left(\frac{\partial U_{tot}(x,\xi)}{\partial \xi}\right)_x = \left(\frac{\partial U_{12}(x,\xi)}{\partial \xi}\right)_x + H_{22}\xi = 0 \qquad [3.65]$$

By combining this equation with relation [3.64], we can eliminate the energy in region 2 and the use of the Hessian matrix (which would be of infinite dimensions), and we obtain:

$$U_{tot}(x,\xi) = U_{11}(x) + U_{12}(x,\xi) - \frac{1}{2}\left(\frac{\partial U_{12}(x,\xi)}{\partial \xi}\right)_x \xi \qquad [3.66]$$

Thus, we are led back to the calculation of the potential energy of region 1 and the energy of interaction between region 1 and region 2 only.

The energy of formation of the defect $U_{def}(x,\xi)$ is, in fact, the difference between the total energy of the perfect crystal $U_{tot}^{pf}(x,\xi)$ and the total energy of the crystal containing the defect $U_{tot}^{df}(x,\xi)$, so:

$$U_{def}(x,\xi) = U_{tot}^{df}(x,\xi) - U_{tot}^{pf}(x,\xi) \qquad [3.67]$$

The next step in the method, therefore, is to look for the values of the x coordinates and the displacements ξ which minimize the energy. In fact, in practice, once the energy is minimal, we recalculate the forces that are exerted between each pair of atoms in region 1 as a function of those distances and displacements, and we minimize the force thus obtained, which is why the method is known as the force equilibrium method.

To calculate the terms $U_{11}(x)$, we consider all the pairs of atoms (or ions) in region 1 and, for each, we calculate the energies resulting from the application of the three potentials: electrostatic, Lennard–Jones attraction and repulsion of the electron clouds.

In order to calculate the term $U_{12}(x,\xi)$, we return to the separation of the two zones 2a and 2b. In region 2a, the forces exerted on the individual ions due to the Lennard–Jones potential and the Coulomb potential are calculated, but to simplify the calculations, experience shows that, in order to take account of region 1, we only need to take account of the forces due to the defect itself. Region 2b is of infinite dimensions, and we only take account of the Coulomb forces. The Coulomb energy of region 1 is replaced by the charge of the defect, and we simply calculate the energy of relaxation induced by that charge. With the computational resources available to us today, region 2b is now considered to be made up of individual points. In their time, Motte and Littleton described the interaction with region 2b by the approximation of a dielectric continuum.

We now have a large number of energies of defects and disorders which have been evaluated. Table 3.9 gives the example of the energies of vacancy formation in atomic solids.

Solid	Kr	Cd	Pb	Zn	Mg	Al	Ag	Cu	Ni
H (kJ / mol)	7.7	38	48	49	56	68	106	120	168

Table 3.9. *Values of some of the energies of formation of a vacancy in atomic solids*

In ionic binary compounds, the orders of magnitude of the defect energies are often a few electron-volts per molecule (a few hundred kilojoules per mole).

4

Solid Solutions and Structure Elements

The modeling of solid solutions, when the solvent and the solute exhibit major behavioral differences, is not easy. The required models quickly become complex. The intervention of structure elements enables us to model such solutions using quasi-chemical equilibria.

In this chapter, we will examine two cases where that modeling enables us to correctly represent the behavior of the poly-constituent solid phase. We will look, in turn, at the study of ionic solid solutions and the fixation of water molecules in the lattices of salts.

4.1. Ionic solid solutions

Ionic solids, unlike metal alloys and molecular solids, exhibit crystalline sites which are very different from each other. Generally speaking, there tends to be a very clear division into anionic sites (which receive anions) and cationic sites (which receive cations). The crystal has two distinct sublattices. In view of this arrangement, the probability of exchange of an ion between a site of one type and site of the opposite type is practically none.

Thus, solid solutions are the solutions of ions which are placed in an interstitial position or in substitution on the sublattice which corresponds to their charge.

As ions are never found alone, the dissolution of an ion will be accompanied by the dissolution of another ion of the opposite sign. If that accompanying ion is already present in the solvent, then the result is the same as if we had simply dissolved the ions not common to the two solids.

If the introduced ions have the same charge as the ions in the solvent, the effect of the dissolution will be very slight. For example, mixed crystals of sodium chloride and potassium chloride can be considered as solutions of Na^+ ions in KCl (or of K^+ ions in NaCl). The resulting solution has all the characteristics of a perfect solution. Indeed, we know that the difference from perfection in a solution is, primarily, due to the exchange energy w_{AB}. In the case of the substitution of K^+ ions by Na^+ ions, this energy is extremely low. This results in large solubility and even in the chosen case, complete miscibility of the two solids.

If, on the other hand, the introduced ion has a different charge to that of one of the ions in the solvent, the consequences are more significant. Such is the case, for example, with the substitution of the Na^+ ions in sodium chloride by calcium ions (Ca^{++}). Two consequences of this substitution are evident.

First, because of the electrostatic forces exerted over a long distance, it is certain that the exchange energy w_{AB} between an ion Na^+ and an ion Ca^{++} is far from negligible. The solution obtained will be far from perfect and solubility will be limited. In this case, it is extremely difficult to model the imperfection of the solution. The easiest way to do so, when the solution is involved in a state of equilibrium, is to use quasi-chemical modeling with structure elements, thus enabling us to consider the solution to be perfect. Certain authors advocate the use of the quasi-chemical approximation, but with the supposition that the solution obeys the Debye and Hückel theory of strong electrolytes[1].

The second consequence of the difference in charge between the initial ion and the introduced ion pertains to the absolute need to respect electrical neutrality in the crystal. This neutrality will be

1 Debye and Hückel's model is presented in Chapter 4 of Volume 2 in this series of books [SOU 15b].

expressed differently depending on whether the host solid is stoichiometric or non-stoichiometric.

This modification of the properties of the initial solid by the dissolution of ions with a different charge has led to the practice of doping, which consists of introducing a foreign element in a solid in a deliberate and controlled manner. We can distinguish between insertional doping, where the foreign element takes up an interstitial position and yields an insertional solid solution, and substitutional doping, where the foreign element takes the place of a normal element in the lattice and yields a substitutional solution.

A dopant may be introduced by reacting the solid in question with another phase (solid, liquid or gaseous) containing the foreign element to be introduced. Let us look at the example of an oxide BO, which we wish to dope with elements A. In order to do so, we can bring the solid BO into contact with the oxide AO of A, at high temperature (to speed up diffusion). An interface reaction will allow A to penetrate into the lattice AO and achieve a homogeneous distribution within it (it is likely that this treatment will also cause the opposite reaction – i.e. the penetration of B into the oxide AO). Another method, which can be used if A is volatile, is to bring the solid BO into contact with the vapor of A at high temperature. A third method consists of using the nitrate of A, dissolved in water, to fill the pores of BO with just the requisite amount of nitrate solution, and then heat the mixture so that the substance is broken down by thermolysis and the elements A penetrate into BO (this technique is known as dry impregnation).

NOTE.– An impurity in a solid solution in the phase in question is simply a dopant, whose presence is both unintended and uncontrolled. Thus, the study of the effects of dissolved impurities in solid phases is identical to the study of dopants.

4.1.1. *Introduction of foreign elements into stoichiometric binary solids*

In order to study the effect of a dopant on a stoichiometric compound, we are going to look at the example of potassium chloride doped with divalent calcium ions. To begin with, we note that the substitutional doping of a stoichiometric compound with an element of the same valence causes no major alteration in the properties of the initial solid.

If we introduce calcium ions, Ca^{2+}, to substitute the potassium ions, K^+, in the potassium chloride, it is necessary to preserve the electrical neutrality of the solid and the ratio of 1:1 between the number of cationic and anionic sites. This is only possible in two scenarios:

– either by the formation of a potassium vacancy for each calcium ion introduced. Figure 4.1(a) gives a two-dimensional (2D) diagrammatic representation of substitution. The introduction reaction is written as:

$$CaCl_2 = Ca_K^{\circ} + 2Cl_{Cl} + V_K^{'} \quad [4R.1]$$

– or by the formation of a chlorine ion in an interstitial position for each calcium ion introduced, as shown in Figure 4.1(b). In this case, the introduction reaction is written thus:

$$CaCl_2 = Ca_K^{\circ} + Cl_{Cl} + Cl_i^{'} \quad [4R.2]$$

In the case of our example, the first possibility is the more likely, because, on the one hand, the chlorine ion is too large to occupy an interstitial position, and on the other hand, pure potassium chloride exhibits Schottky disorder, with vacancies of potassium and chlorine. Thus, we can see that the introduction of calcium alters the ratio between the anionic vacancies and the cationic vacancies in potassium chloride.

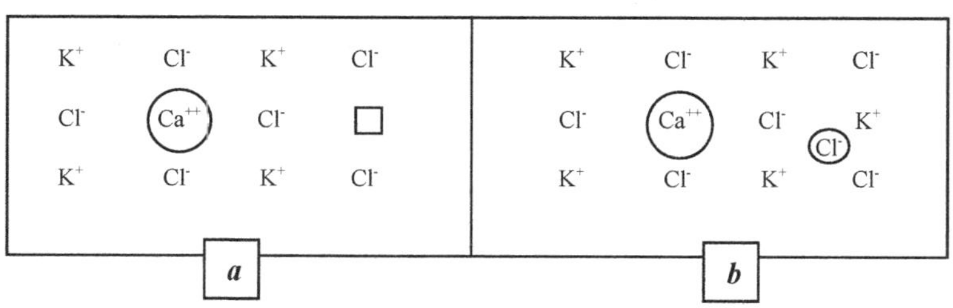

Figure 4.1. *Diagram of potassium chloride a) in the pure state and b) doped with calcium ions*

We can look at the problem in quantitative terms. In order to do this, we write the law of mass action for the creation of Schottky disorder:

$$0 = \mathrm{V}_{\mathrm{K}}^{'} + \mathrm{V}_{\mathrm{Cl}}^{\circ} \qquad [4R.3]$$

Thus:

$$\left[\mathrm{V}_{\mathrm{K}}^{'}\right]\left[\mathrm{V}_{\mathrm{Cl}}^{\circ}\right] = K_S \qquad [4.1]$$

By finding the balance of the different types of sites, we can write:

$$\left[\mathrm{V}_{\mathrm{Cl}}^{\circ}\right] + \left[\mathrm{Ca}_K^{\circ}\right] = \left[\mathrm{V}_{\mathrm{K}}^{'}\right], \text{ where } \left[\mathrm{Ca}_K^{\circ}\right] = c \qquad [4.2]$$

We will now examine the effect of doping in the significant case characterized by the Brouwer domain:

$$\left[\mathrm{V}_{\mathrm{K}}^{'}\right] = \left[\mathrm{Ca}_K^{\circ}\right] \qquad [4.3]$$

On the basis of relations [4.1] and [4.3], we deduce:

$$\left[\mathrm{V}_{\mathrm{K}}^{'}\right] = c \text{ and } \left[\mathrm{V}_{\mathrm{Cl}}^{\circ}\right] = K_S / c \qquad [4.4]$$

These relations are valid if the following condition is fulfilled:

$$c > \left[\mathrm{V}_{\mathrm{Cl}}^{\circ}\right] \text{ so } c > K_s^{1/2} \qquad [4.5]$$

Figure 4.2 shows the variations of the logarithms for the vacancy concentrations as a function of the logarithm for calcium concentration with the two domains, with the second domain fulfilling condition [4.5], whereas for low values of c, we find ourselves with the case of practically pure potassium chloride with equal numbers of chlorine vacancies and potassium vacancies.

Thus, the doping of potassium chloride with cations whose valence is greater than 1 results in an increase in the concentration of cationic vacancies and a decrease in the concentration of anionic vacancies.

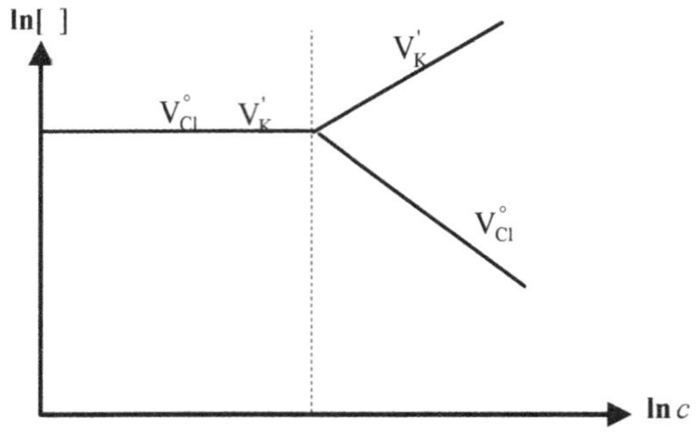

Figure 4.2. *Concentrations of vacancies in calcium-doped potassium chloride*

The calculation can be carried out for a cationic impurity with valence *Z'*, substituting for cations of valence *Z* for the different types of disorder of binary stoichiometric solids. Table 4.1 gives the results obtained in the cationic doping of a metaloxide MO, of Schottky or Frenkel, on M.

Type of solid	*Z'>Z*		*Z'<Z*	
Schottky	$[V_M]$ increases	$[V_O]$ decreases	$[V_M]$ decreases	$[V_O]$ increases
Frenkel	$[M_i]$ decreases	$[V_M]$ increases	$[M_i]$ increases	$[V_M]$ decreases

Table 4.1. *Influences of dopants on stoichiometric binary solids*

We see the same result as for Schottky potassium chloride doped with calcium (*Z'* > *Z*).

4.1.2. *Influence of foreign elements introduced into a non-stoichiometric binary solid*

The addition of a foreign element whose valence is different from those of the basic components will enable us to control the electronic deviation from the Wagner disorder, for example. We will illustrate this point by the introduction of lithium into iron(II) oxide in cationic substitution. The valence Z of the basic element being replaced is +2; that of the foreign substitute is Z' = +1. Under normal conditions, this iron oxide contains cationic vacancies, compensated by electron holes trapped on Fe^{+2} ions in the lattice, transforming them into Fe^{+3} ions. Figure 4.3(a) offers an illustration of pure iron oxide.

The introduction of lithium into vacancies in iron has the effect of modifying the number of Fe^{+3} ions (in actual fact, the number of electron holes), the introduction reaction is as follows, if we use lithium vapor:

$$\mathrm{Li}\,(gas) + \mathrm{V}''_{\mathrm{Fe}} + \mathrm{Fe}^{\circ}_{\mathrm{Fe}} = \mathrm{Li}'_{\mathrm{Fe}} + \mathrm{Fe}_{\mathrm{Fe}} \qquad [4R.4]$$

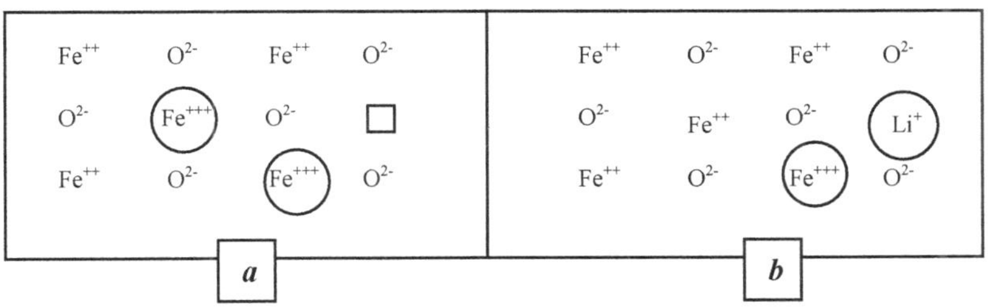

Figure 4.3. *Diagram of iron oxide a) in the pure state and b) doped with lithium*

Figure 4.3(b) illustrates the new situation. Li^{+} ions are placed to substitute the Fe^{2+} ions in the vacancies present in the pure solid. If the number of lithium ions introduced is sufficient, there are no longer enough iron vacancies to accommodate them, and we then see that the lithium ions introduced are placed in an interstitial position, which will lead to a further reduction in the number of Fe^{3+} ions.

In order to quantitatively appreciate the effect of doping, let us examine the first case of lithium ions introduced in substitution of Zn^{2+} ions in zinc oxide, at equilibrium with the oxygen. Zinc oxide exhibits overstoichiometry of zinc, which results in the presence of zinc ions in an interstitial position and free electrons. The equilibrium with oxygen is written as follows:

$$\mathrm{Zn_i^{\circ\circ}} + 2\mathrm{e'} + 1/2\mathrm{O_2} = \mathrm{Zn_{Zn}} + \mathrm{O_O} \qquad [4R.5]$$

The concentration of lithium is supposed to be fixed at the value c:

$$\left[\mathrm{Li}'_{Fe}\right] = c \qquad [4.6]$$

Application of the law of mass action to the equilibrium relation [4R.5] gives us:

$$K = \frac{1}{[\mathrm{e'}]\left[\mathrm{Zn_i^{\circ\circ}}\right]P^{1/2}} \qquad [4.7]$$

In light of the condition imposed by electrical neutrality, which is of the form:

$$2\left[\mathrm{Zn_i^{\circ\circ}}\right] = \left[\mathrm{Li}'_{\mathrm{Zn}}\right] + [\mathrm{e'}] \qquad [4.8]$$

We are led to define two Brouwer regions:

– region 1 characterized by: $2\left[\mathrm{Zn_i^{\circ\circ}}\right] = [\mathrm{e'}]$, so $[\mathrm{e'}] \ll c$ [4.9]

– region 2 characterized by: $2\left[\mathrm{Zn_i^{\circ\circ}}\right] = [\mathrm{e'}]$, so $[\mathrm{e'}] \gg c$ [4.10]

Table 4.2 gives the concentrations of vacancies for the two regions. The boundary between them obeys the condition:

$$[\mathrm{e'}] = c \text{, so } Pc^6 = \frac{4}{K^2} \qquad [4.11]$$

Elements	Region 1 $Pc^6 = \frac{4}{K^2}$	Region 2 $Pc^6 = \frac{4}{K^2}$
$[\mathrm{e'}]$	$\sqrt{\frac{2}{Kc}}\frac{1}{P^{1/4}}$	$\left(\frac{2}{K}\right)^{1/3}\frac{1}{P^{1/6}}$
$\left[\mathrm{Zn_i^{\circ\circ}}\right]$	$\frac{c}{2}$	$\frac{1}{2}\left(\frac{2}{K}\right)^{1/3}\frac{1}{P^{1/6}}$

Table 4.2. *Concentrations of vacancies in lithium-doped zinc oxide*

The effect of the dopant is only apparent in region 1. Figure 4.4(a) shows the Kröger–Vink diagram as a function of the lithium concentration at a given oxygen pressure.

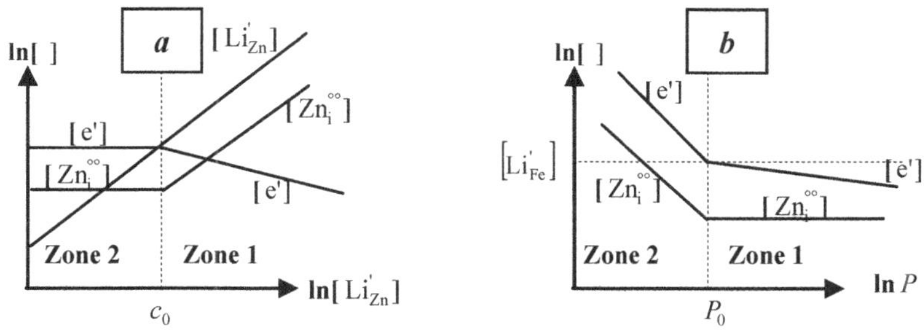

Figure 4.4. *Kröger-Vink diagrams for lithium-doped zinc oxide*

In particular, in zone 1 (on the right of the diagram), we note that the concentration of electrons is inversely proportional to the square root of the lithium concentration.

Figure 4.4(b) shows the diagram as a function of the oxygen pressure at a given concentration of lithium introduced. We can see that at sufficiently high pressure (zone 1), the concentration of interstitial zinc ions becomes independent of the oxygen pressure.

If the lithium were replaced by a trivalent cation – i.e. one whose valence Z' is greater than the valence Z of the cation being substituted – we would see variations in the opposite direction.

Table 4.3 shows the direction of variation of the concentration of charge carriers in a semi-conductor, as the result of the substitutional doping of the cation of valence Z with a dopant of valence Z'.

Type of oxide	Z'>Z	Z'<Z
p	Number of charge carriers decreases	Number of charge carriers increases
n	Number of charge carriers increases	Number of charge carriers decreases

Table 4.3. *Influence of a dopant on a Wagner semi-conductor*

NOTE.– It is easy to show that doping with cations in an interstitial position, regardless of the charge on the ion introduced, yields the same qualitative results as substitutional doping with a cation of higher valence.

We can easily show that doping with an anion leads to qualitative results opposite to those obtained with cations.

NOTE.– Induced valence. In certain cases, the foreign atom can assume several degrees of oxidation, and its charge adapts to that of the normal elements of the lattice, which then produce no effect. We say that the valence of the dopant is induced by the lattice.

4.2. Thermodynamics of equilibria between water vapor and saline hydrates: non-stoichiometric hydrates

Numerous salts are capable of fixing water. This introduced water may either remain in molecular form (H_2O) or be dissociated, with the then appearing as solutions of $OH^-$and H^+ ions. We are going to look at the molecular dissolution of water in salts.

The equilibrium between two hydrated salts and water vapor can be written thus:

$$S,nH_2O + q\ H_2O = S,(n+p)H_2O \qquad [4R.6]$$

The numbers n and q are two positive numbers, although the value of n may actually be zero.

We will see that this overall equilibrium can be refined, because our formulation does not specify whether the two hydrated forms with n and $n + q$ water molecules form a single solid phase or constitute two distinct solid phases.

4.2.1. *Experimental demonstration of non-stoichiometry of a hydrate*

When we look at the equilibrium states of such a system, we usually use isothermal- and isobaric-thermogravimetry at a controlled water vapor pressure.

We will use ε to denote the total amount of water retained by the solid per mole of salt S. The amount of water involved in the equilibrium reaction is $\varepsilon - n$.

The maximum value of ε is ε_{lim}, such that:

$$\varepsilon_{\text{lim}} = n + p \text{ and therefore } \varepsilon_{\text{lim}} - n = p \qquad [4.12]$$

Experimentally, we can see that two forms of curves are obtained for to equilibrium isotherms.

The first shape of the curve, represented in Figure 4.5(a), is characterized by the presence of two steps which correspond to the compositions of the two hydrates with n and $n + q$ moles of water, separated by a vertical at a definite pressure. If we change the temperature, the shape of the curve remains the same; only the vertical separating the two platforms is placed at a different value of the pressure.

The second shape of the curve, shown in Figure 4.5(b), is characterized by a constant variation of the water content of the whole solid as the pressure varies. At a different temperature, a new curve exhibiting the same characteristic is obtained.

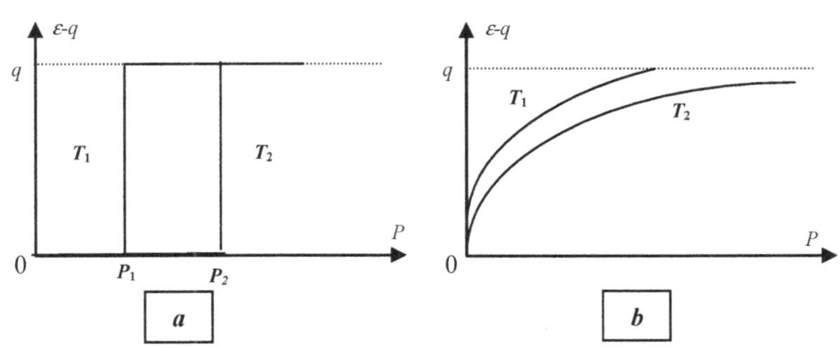

Figure 4.5. *Equilibrium isotherm between water vapor and a) a stoichimetric hydrate and b) a non-stoichiometric hydrate*

The first experimental result thus obtained is, in fact, the variance of the system. Indeed, in the case of Figure 4.5(a), we see that if the temperature remains static, equilibrium between the two compositions only occurs at a given pressure, represented by the vertical, and therefore the variance of the system is 1. However, in the case of Figure 4.5(b), the water content of the solid varies continuously with the pressure at a given temperature, and each point depends on the temperature, so the variance of the system is 2.

Let us look again at Gibbs' phase rule for these systems. In both cases, the number of independent components is 2 (salt and water), and the number of external parameters taken into account is also 2 (pressure and temperature). In both systems, there is a gaseous phase, and if we let φ_s denote the number of solid phases, then Gibbs' law is written as follows:

$$v = 2 + 2 - 1 - \varphi_s \text{ and therefore } \varphi_s = 3 - v \qquad [4.13]$$

We can use this, in both cases, to deduce the number of solid phases mentioned in Table 4.4.

Figure	Variance	Number of phases	Solid phases
(a)	1	3	2
(b)	2	2	1

Table 4.4. *Number of solid phases in the equilibrium of a hydrated solid and water vapor*

If the variance is 1, as is the case in Figure 4.5(a), there are two solid phases at equilibrium: the two hydrates, each with a very specific composition of n and $n + p$ water molecules.

If the variance is 2, we say that we have a divariant hydrate: the two hydrate forms with n and $n + p$ water molecules are merely two extreme forms of the single hydrate phase, whose composition varies continuously with the pressure and temperature. Thus, it is a solid solution or, which is equivalent, a non-stoichiometric hydrate. The water contained in the solid is sometimes referred to as zeolitic because zeolites constitute a family of solids which have that property shared by other hydrates noted in the existing literature.

4.2.2. *Equilibria between stoichiometric hydrates*

In the equilibrium reaction [4R.6], the two hydrates constitute two solid phases, the law of mass action applied to that equilibrium is written as:

$$K_6 = P = K_6^0 \exp\left(-\frac{\Delta_6 H}{\mathrm{R}T}\right) \qquad [4.14]$$

This is expressed by the fact that at a given temperature, the equilibrium water vapor is fixed.

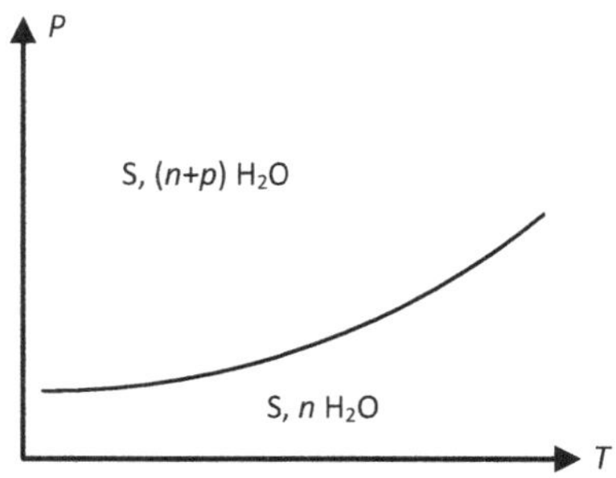

Figure 4.6. *Pressure–temperature diagram for a stoichiometric hydrate*

Figure 4.6 shows the curve for this equilibrium in the pressure–temperature diagram. On the left of the curve, we see the domain of stability of the superior hydrate ($q = n + p$), and on the right of the curve, we see the domain of stability of the inferior hydrate (n). The value of n may be 0, in which case we have the anhydrous form. The enthalpy associated with the transformation is generally positive (endothermic dehydration).

4.2.3. *Equilibrium reactions in non-stoichiometric hydrates*

We will now model divariant equilibria by non-stoichiometry of the hydrates. Various models are envisaged, and we interpret different examples.

4.2.3.1. *Definition of models*

As is the case with the different solid compounds found in this volume, we will model non-stoichiometry of the hydrates using quasi-chemistry of structure elements. This approach, however, presents a number of difficulties because these hydrates are relatively-complex solids, with at least three main components: the anion (which itself is usually complex), the cation and water. In cases where the salt can accept several successive limited hydrates, even the water molecules are not all equivalent in terms of the sites they occupy and the energy of their bond to the lattice. In order to simplify the model, we will use a pseudo-binary approach, considering the hydrated salt to be formed of two main components – one of which is the water involved in the equilibrium in question (p molecules per salt molecule) and the other is the skeletal structure of the salt, involving the anhydrous part, and possibly the n water molecules of inferior hydrates not involved in the equilibrium.

For example, we will look at isothermal curves. Thus, we can disregard the point defects in the skeleton, which will remain more or less constant in nature and in number. Hence, we are only interested in the structure elements relating to water, and we distinguish:

– the water molecules in a normal position in the lattice, written as $<H_2O>$ or $<H_2O>_q$, if those elements are associated q with q;

– the water molecules in an interstitial position written as (H_2O) or $(H_2O)_q$;

– the water vacancies, written as $<>$ or $<>_q$;

– the vacant interstitial positions, written as $(\ )$ or $(\ )_q$;

– the non-localized water molecules, which are mobile in the solid, written as $<<H_2O>>$.

The water in the gaseous state will be written as $[H_2O]$.

We will make the hypothesis of the predominant defect in a given hydrate, supposing that water molecules are never to be found on a site normally occupied by the skeleton.

We will look at two limit cases of non-stoichiometry of the hydrates:

– the water molecules are all delocalized and free to move about in the lattice (e.g. in channels);

– the water molecules are localized on specific sites in the solid, which leads us to two possibilities; substoichiometric hydrates which exhibit water vacancies and overstoichiometric hydrates which have water molecules in an interstitial position.

4.2.3.2. *Non-stoichiometric hydrates with mobile water molecules*

As the water molecules do not occupy specific crystallographic positions, we define the model in quasi-chemical form:

$$[H_2O] = <<H_2O>> \qquad [4R.7]$$

If we assign the index 1 to all the values relating to the skeleton, the index 2 to those relating to the water molecules in the solid and the index G to the values relating to the gases, it is easy to show that all of the equilibrium positions are given by the relation:

$$\left(S_G - \overline{S}_2\right)\mathrm{d}T - \left(V_G - \overline{V}_2\right)\mathrm{d}P + \left(\frac{\partial \mu_2}{\partial \varepsilon}\right)_{P,T} \mathrm{d}\varepsilon = 0 \qquad [4.15]$$

From this expression, we can obtain various curves. In order to simplify the calculations, we set $\overline{V}_2 << V_G = \dfrac{\mathrm{R}T}{P}$. We define the enthalpy by: $\Delta_7 H = H_G - \overline{H}_2$. In particular, we can extract the isothermal curves defined by $\mathrm{d}T = 0$:

$$\left(\frac{\mathrm{d}\ln P}{\mathrm{d}\varepsilon}\right)_T = \frac{1}{\mathrm{R}T}\left(\frac{\partial \mu_2}{\partial \varepsilon}\right)_{P,T,n_1=1} \qquad [4.16]$$

However, the chemical potential of the water in the solid is of the form:

$$\mu_2 = \mu_2^0 + RT \ln \gamma_2 x_2 \qquad [4.17]$$

and it is easy to show that the molar fraction of the water in the solid is:

$$x_2 = \frac{\varepsilon - n}{1 + \varepsilon - n} \qquad [4.18]$$

Hence, we obtain the following expression for the isotherm:

$$\varepsilon - n = \frac{P}{\gamma_2 K_7 - P} \qquad [4.19]$$

Curve 1 in Figure 4.7 shows the shape of that isotherm in the case of ideal behavior of water in the solid ($\gamma_2 = 1$). We can see that the concavity of that curve is turned upward, except near to saturation, which is not taken into account by the model because, near to the saturation point, the water molecules lose their mobility.

We can take account of the interactions between the water molecules and the solid by way of the coefficient γ_2. We can make the model of the solution more complex by decreasing the mobility of the water, and we see a tendency toward a curve with down-turned concavity (curve 4 in Figure 4.7) which, as we will see later on (section 4.2.3.3), characterizes models where the water molecules are localized.

We can also take account of the interactions between the water molecules by considering molecules associated q with q. The equilibrium is then written as:

$$q[H_2O] = \langle\langle H_2O \rangle\rangle_q \qquad [4R.8]$$

The expression of the isotherm then becomes:

$$\varepsilon - n = \frac{qP^q}{\gamma_2 K_8 - P^q} \qquad [4.20]$$

We then obtain a curve similar to that shown by curve 2 in Figure 4.7. When the value of q increases, the curve tends toward the straight line 3 in Figure 4.7.

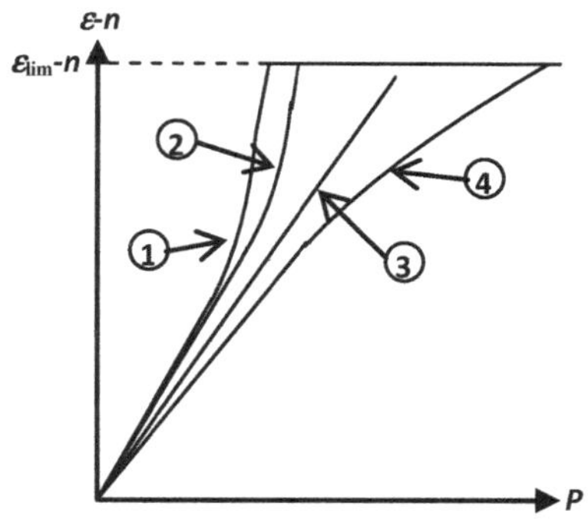

Figure 4.7. *Equilibrium isotherms between water vapor and a non-stoichiometric hydrate with non-localized water molecules*

4.2.3.3. *Non-stoichiometric hydrates with localized water molecules*

In this category of models, the water molecules are localized. They may either be in their normal position (in which case we would have substoichiometric hydrates of the limited hydrate S,($n + p$) H_2O), or in an interstitial position, in which case we are dealing with an overstoichiometric hydrate of the inferior hydrate S, nH_2O.

4.2.3.3.1. Substoichiometric hydrates

In this model, the water molecules occupy specific crystallographic positions, and some of those positions are not occupied, thus constituting water vacancies. In the most usual case of associated defects q and q, the quasi-chemical equilibrium is written as follows:

$$q[H_2O] + \langle\rangle_q = \langle H_2O\rangle_q \qquad [4R.9]$$

If the indices 1 and 2 are assigned, respectively, to the relative values of the water vacancies and the normal elements of water, it is easy to show that the molar fractions of those entities are given by:

$$x_1 = \frac{n+p-\varepsilon}{p+q} \text{ and } x_2 = \frac{\varepsilon - n}{p+q} \qquad [4.21]$$

We can easily show that the set of equilibrium states satisfies the relation:

$$\frac{\Delta_9 H^{\circ}}{T}\mathrm{d}T + \mathrm{R}T\ln P^{q} + \left(\frac{\partial\mu_1}{\partial\varepsilon} - \frac{\partial\mu_2}{\partial\varepsilon}\right)\mathrm{d}\varepsilon = 0 \qquad [4.22]$$

From this, we deduce the expression of the isotherm by making $\mathrm{d}T = 0$:

$$\frac{\mathrm{d}\ln P^{q}}{\mathrm{d}\varepsilon} = \frac{1}{\mathrm{R}T}\left(\frac{\partial\mu_1}{\partial\varepsilon} - \frac{\partial\mu_2}{\partial\varepsilon}\right) \qquad [4.23]$$

If we take account of the expressions of the chemical potentials, which are of the form of equation [4.17], the expression of the isotherm becomes:

$$\varepsilon = n + \frac{p\dfrac{\gamma_1}{\gamma_2}P^{q}}{K_9 + \dfrac{\gamma_1}{\gamma_2}P^{q}} \qquad [4.24]$$

The simplest form of this isotherm is obtained for an ideal solution ($\gamma_1 = \gamma_2 = 1$) and $q = 1$, so:

$$\varepsilon = n + \frac{pP}{K_9 + P} \qquad [4.25]$$

This expression is illustrated by curve 1 in Figure 4.8. We can see that the concavity of the curve is now turned downward, unlike with curve 1 in Figure 4.7, which pertains to mobile water molecules.

Curve 2 in Figure 4.8 is the isotherm given by relation [4.22] for ideal solutions where $q > 1$. In this case, if we know the value of q, the position of the point of inflection can be used to calculate the equilibrium constant, and vice versa.

A more refined model can be used, with activity coefficients which are functions of the composition; this is tantamount to increasing the interactions between the water molecules. Thus, eventually, we have to gradually move from the isotherm with $q = 1$ (curve 1) to the isotherm $q > 1$ (curve 2). This is what we see when we choose the model of a strictly-regular solution, for example. The system can continue up to the point of demixing.

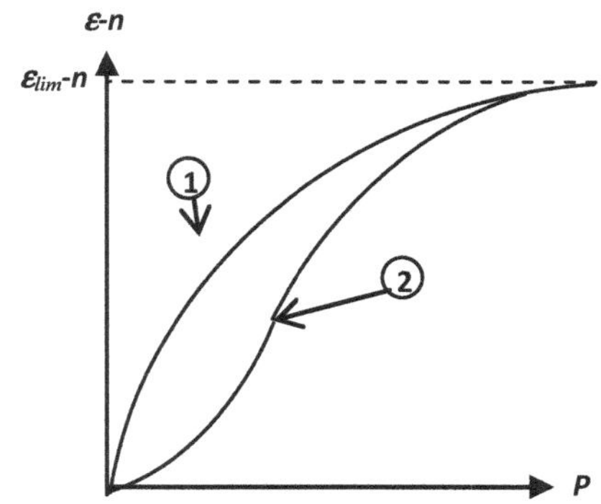

Figure 4.8. *Isothermal curves showing the equilibrium between water vapor and a non-stoichiometric hydrate with localized water molecules*

4.2.3.3.2. Overstoichiometric hydrates

This time, the water molecules are in an interstitial position. The corresponding equilibrium is written thus:

$$q[H_2O] + (\ )_q = (H_2O)_q \qquad [4R.10]$$

The calculation tells us that this view of things also yields expression [4.24] for the isotherm. In fact, this means that a zone of divariance with localized water molecules can be viewed either as a substoichiometric hydrate in relation to a superior hydrate, or as an overstoichiometric hydrate in relation to an inferior hydrate (or an anhydrous salt). The two models are absolutely equivalent.

4.2.4. The limits of the domains of divariance

In the pressure–temperature space, the domain of divariance is generally limited. In most cases, a phenomenon, represented by an equilibrium reaction, limits the domain by superposing itself on the non-stoichiometry equilibrium. We can distinguish two categories of phenomena:

– those which involve water in the vaporous state, such as the liquefaction of water, which we write as:

$$[H_2O] = H_2O_{\text{liquid}} \qquad [4R.11]$$

– those which only involve water vapor, such as the precipitation of the inferior hydrate into a new crystallographic structure – i.e. a new phase. We can represent this precipitation by the equilibrium reaction:

$$\langle\langle S, nH_2O \rangle\rangle_{\text{Phase 1}} = \langle S, nH_2O \rangle_{\text{Phase 2}} \qquad [4R.12]$$

with "phase 1" being the stoichiometric phase and "phase 2" being the non-stoichiometric phase.

By way of example, we will look at this final case. The limit of the domain of divariance is characterized by the equilibrium [4R.9], where $q = 1$ and in ideal solution conditions, i.e. represented by the isotherm [4.25]. The new equilibrium can be written, in structure elements, as the simple disappearance of the vacancies (without conservation of the sites following the phase change), so:

$$V_{H_2O} = 0 \qquad [4R.13]$$

In order to examine the limit of the divariance zone in the $P(T)$ diagram, we superimpose the equilibrium reactions [4R.9] and [4R.13]. By application of the law of mass action to the equilibrium relation [4R.13], we find that the molar fraction of the vacancies is:

$$x_1 = \frac{1}{K_{13}} = \exp\frac{\Delta_{13}G^\circ}{RT} \qquad [4.26]$$

and, by substituting that value back into the expression of the isostere, obtained by making $d\varepsilon = 0$ in expression [4.15] and taking account of the fact that the sum of the fractions of sites of the water molecules and the vacancies is equal to 1, we obtain:

$$P = \exp-\left(\frac{\Delta_0 G^\circ}{RT}\right)\left[\exp-\left(\frac{\Delta_{13} G^\circ}{RT}\right)-1\right] \qquad [4.27]$$

This limit is represented by the curve marked "1" in Figure 4.9, and separates the phase 1 domain from the domain of divariance in the non-stoichiometric phase 2. The introduction of the new equilibrium lowers the variance of the system to 1. The curves labeled "2" in Figure 4.9 represent the curves of isocomposition (isosteres) in the non-stoichiometric phase.

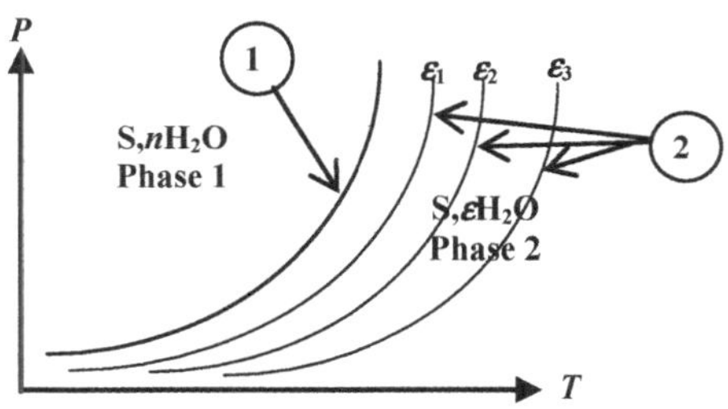

Figure 4.9. *Domain of divariance limited by precipitation of the inferior hydrate into a new solid phase*

NOTE.– Phase 1 itself may be another zone of divariance if the inferior hydrate in its new structure is, itself, non-stoichiometric, in which case, two domains, separated by the curve representing the separation of the two phases, can be described as two phases which represent the hydrate with n water molecules – one substoichiometric (phase 1) and the other overstoichiometric in relation to n (phase 2).

APPENDICES

Appendix 1

The Lagrange Multiplier Method

A.1.1. Statement of the problem

We wish to find the value of the extremum (maximum or minimum) of a function f of several variables $x_1, x_2, \dots, x_n$.

If the variables undergo a slight variation from $x_1 + \delta x_1$ to $x_i + \delta x_i$, the function varies from f to $f + \delta f$, such that:

$$\delta f = \sum_{i=1}^{n} \left(\frac{\partial f}{\partial x_i} \right) \delta x_i$$

For an extremum, $\delta f = 0$, and therefore:

$$\sum_{i=1}^{n} \left(\frac{\partial f}{\partial x_i} \right) \delta x_i = 0 \qquad \text{[A.1.1]}$$

If all the variables are independent, this equation can be solved by zeroing each term in the sum:

$$\left(\frac{\partial f}{\partial x_i} \right) = 0 \qquad \forall x_i$$

The problem which interests us here is how to find the extremum if all the variables x_i are not independent, but instead there are relations between them, which are called "constraints"; in this case, the above solution is no longer valid. We then use the Lagrange multiplier method.

A.1.2. Solution by the multiplier method

Let us suppose, in order to illustrate the method, that there is a constraint which exists between the variables x_i such that:

$$g(x_i) = 0$$

The constraint g is always true, and the value of g remains unchanged when the values of the variables x_i vary, so:

$$\delta g(x_i) = \sum_{i=1}^{n} \left(\frac{\partial g}{\partial x_i} \right) \delta x_i = 0$$

Because δg is null, we can multiply it by an arbitrary parameter λ and add it to equation [A.1.1], in which case we obtain:

$$\sum_{i=1}^{n} \left\{ \left(\frac{\partial f}{\partial x_i} \right) + \lambda \left(\frac{\partial g}{\partial x_i} \right) \right\} \delta x_i = 0 \qquad \text{[A.1.2]}$$

Equation [A.1.2] can be solved for one of the variables – e.g. for $i = n$, as a function of the other variables ($i = 1, 2, \ldots, n-1$), which are independent. For the time being, λ is arbitrary, but we can perfectly well choose its value so that the term in x_n in equation [A.1.2] is zero – that is we can choose λ in such a way that:

$$\left(\frac{\partial f}{\partial x_n} \right) + \lambda \left(\frac{\partial g}{\partial x_n} \right) = 0 \qquad \text{[A.1.3]}$$

Thus, equation [A.1.2] becomes:

$$\sum_{i=1}^{n-1} \left\{ \left(\frac{\partial f}{\partial x_i} \right) + \lambda \left(\frac{\partial g}{\partial x_i} \right) \right\} \delta x_i = 0$$

Now, the n–1 variables are independent and the solution is:

$$\left(\frac{\partial f}{\partial x_i}\right)+\lambda\left(\frac{\partial g}{\partial x_i}\right)=0 \quad i=1,2,\ \ldots,\ n$$

However, equation [A.1.3] has exactly the same form, and thus the maximum or minimum of f can be found by solving the system:

$$\left(\frac{\partial f}{\partial x_i}\right)+\lambda\left(\frac{\partial g}{\partial x_i}\right)=0 \quad i=1,2,\ \ldots,\ n$$

If there are multiple constraints, then we introduce into [A.1.2] as many multipliers as there are constraints, and deduce the same number of equations in the form of [A.1.3].

A.1.3. Determination of the values of the multipliers

There are two possible methods to determine the value of the arbitrary constant λ.

The first method is to solve equation [A.1.3] rather than injecting it into the minimization process.

The second method is to keep the multiplier λ as an unknown until we can deduce a property whose value is known. It is this method which is often used in statistical thermodynamics to determine the multiplier $\beta = 1/\mathrm{k_B}T$.

In the case of several multipliers, the above methods can be mixed, with some of the constants being determined by the first method, and the others by the second method. We see such mixing in determining the two constants α and β in statistical thermodynamics (see sections 4.2.2, 4.2.3, 4.5.3 and 4.5.4).

Appendix 2

Solving Schrödinger's Equation

The Schrödinger equation governs the wave mechanics to be used instead of classical mechanics for atomic environments. For a set of particles i and time-independent systems, this equation is of the form:

$$-\frac{\hbar^2}{2}\sum_i\left[\frac{1}{m_i}\left(\frac{\partial^2\Phi}{\partial x^2}+\frac{\partial^2\Phi}{\partial x^2}+\frac{\partial^2\Phi}{\partial z^2}\right)\right]+V\Phi=W\Phi$$

In this equation, V is the potential energy and W is the total energy of the system.

Except in very rare cases, Schrödinger's equation is too complicated to be solved exactly, and we often merely content ourselves with finding an approximate solution by various methods; one of the most widely used such methods is the variation method.

We choose a normed function Φ dependent on various parameters – obtained, for example, by linear combination of linearly-independent normed functions $\phi_1, \phi_2 ... \phi_i ... \phi_n$ (this is known as the linear state superposition principle):

$$\Phi = \sum_n \lambda_i \phi_i$$

There is no reason for Φ to be a solution to Schrödinger's equation. Also, the average energy $\overline{W}$ of the system in that hypothetical state is certainly greater than the energy W_0, which is as low as possible and corresponds to the system's base state.

For the sake of simplicity of the notation, let us write:

$$S_{i,j} = \int_n \phi_i^* \phi_j \, \mathrm{d}\tau \text{ and } H_{i,j} = \int_n \phi_i^* \, \mathrm{H} \phi_j \, \mathrm{d}\tau$$

By norming Φ, we are able to write:

$$N = \int_n \Phi^* \Phi \, \mathrm{d}\tau = \int_n \sum_i \left(\lambda_i^* \phi_i^* \right) . \sum_i \left(\lambda_i \phi_i \right) \mathrm{d}\tau = \sum_{i,j} \lambda_i^* \lambda_j S_{i,j} = 1$$

The average value of the energy is:

$$\overline{W} = \int_n \Phi^* \, \mathrm{H} \Phi \, \mathrm{d}\tau = \int_n \sum_i \left(\lambda_i^* \phi_i^* \right) \mathrm{H} \sum_i \left(\lambda_i \phi_i \right) \mathrm{d}\tau = \sum_{i,j} \lambda_i^* \lambda_j H_{i,j}$$

We are going to determine the minimum value of the average energy, which will be an approximate value of W_0, by choosing appropriate values for the coefficients λ_l, while keeping $N = 1$. In order to do so, we employ the Lagrange multiplier method (see Appendix 1).

Thus, we obtain a system of n homogeneous linear equations such that:

$$\sum_j \lambda_j^* \left(H_{i,j} - W S_{i,j} \right) = 0 \qquad \text{[A.2.1]}$$

This system can only accept a solution (other than $\lambda_j = 0$ for all values of j) if its determinant is zero, which we can write in shortened form as:

$$\left| H_{i,j} - W S_{i,j} \right| = 0 \qquad \text{[A.2.2]}$$

By zeroing the determinant, we find an equation, called the secular equation, of degree n in W. The smallest root of this equation gives us an approximate value for W_0. J.K.L. MacDonald showed that the other roots constituted approximate values of the energy in the different excited states of the system.

With each root W_k of the secular equation, the system shown in equation [A.2.1] matches a certain corresponding function Φ_k.

It appears that the error committed by replacing a value of the energy with the corresponding root W_k of the secular equation is less than $\sqrt{D_k - W_k^2}$, where D_k and W_k denote the integrals:

$$D_k = \int_n \left(\mathrm{H}\Phi_k^*\right)\left(\mathrm{H}\Phi_k\right).\mathrm{d}\tau \text{ and } W_k = \int_n \Phi_k^* \,\mathrm{H}\Phi_k.\mathrm{d}\tau$$

Bibliography

[ATK 90] ATKINS P.W., *Physical Chemistry*, Oxford University Press, Oxford, 1990.

[BEN 12] BEN-NAIM A., *Entropy and the Second Law*, World Scientific Publishing Company, New Jersey, 2012.

[BER 97] BERNACHE-ASSOLLANT D., COURNIL M., *Equilibres en solution*, Masson, Paris, 1997.

[BES 84] BESSON J., *Précis de thermodynamique et de cinétique électrochimiques*, Ellipses, Paris, 1984.

[CHR 75] CHRISTIAN J.W., *Theory of Transformations in Metals and Alloys*, 2nd ed., Pergamon Press, Oxford, 1975.

[DEF 51] DEFAY R., PRIGOGINE I., *Tension superficielle et adsorption*, Edition Desoer, Liège, 1951.

[DEM 73] DEMANGE H., GERMAIN G., NOTIN M., *Comprendre et appliquer la thermodynamique*, Masson, Paris, 1973.

[DEN 68] DENBIGH K., *The Principles of Chemical Equilibrium*, Cambridge University Press, Cambridge, 1968.

[DES 10] DESRE P., HODAJ F., *Thermodynamique des matériaux. Equilibres de phase etmétastabilité*, EDP Sciences, Paris, 2010.

[EMS 51] EMSCHWILLER G., *La chimie physique*, PUF, Paris, 1951.

[FOW 49] FOWLER R., GUGGENHEIM E.A., *Statistical Thermodynamics*, Cambridge University Press, London, 1949.

[GER 59] GERMAIN J.E., *Catalyse hétérogène*, Dunod, Paris, 1959.

[GIL 63] GILBERT J., *Chimie-Physique I. Atomistique et liaison chimique*, Masson, Paris, 1963.

[GLA 41] GLASSTONE S.N., LAIDLER K.J., EYRING H.W., *The Theory of Rate Processes*, McGraw-Hill, New York, 1941.

[GOK 96] GOKCEN N.A., REDDY R.G., *Thermodynamics*, Plenum Press, New York, 1996.

[GRE 99] GREINER W., *Mécanique quantique. Une introduction*, Springer, Berlin, 1999.

[GUG 32] GUGGENHEIM E.A., "On the statistical mechanics of dilute and of perfect solutions", *Proc. Roy. Soc. A*, vol. 135, p. 181, 1932.

[HIL 56] HILL T.L., *Statistical Mechanics*, McGraw-Hill, New York, 1956.

[INF 06] INFELTA P., GRAETZEL M., *Thermodynamique, principes et applications*, Brown Walker Press, Boca Raton, FL, 2006.

[KLO 08] KLOTZ I.M., ROSENBERG R.W., *Chemical Thermodynamics: Basic Concepts and Methods*, 7th ed., Wiley & Sons, New York, 2008.

[KRÖ 74] KRÖGER F.A., *The Chemistry of Imperfect Crystals*, North-Holland Publishing Company, Amsterdam, 1974.

[LAL 06] LALAUZE R., *Physico-chimie des interfaces solide-gaz 1*, Hermes Science-Lavoisier, Paris, 2006.

[LEG 05] LEGORREC B., Thermodynamique et cinétique électrochimique expérimentale, available at: www.minatec.grenoble-inp.fr/Cours-A.pdf, 2005.

[MAR 95] MARCHAND A., PACAULT A., *La thermodynamique mot à mot*, De Boeck University, Brussels, 1995.

[MCQ 99] MCQUARRIEN D.A., SIMON J.D., *Molecular Thermodynamics*, University Science Books, Sausalito, 1999.

[MEZ 46] MEZGER E., "Loi de variation de la tension superficielle avec la temperature", *J. Phys. Radium*, vol. 7, no. 10, pp. 303–309, 1946.

[MOT 38] MOTT N.F., LITTLETON M.J., "Conduction in polar crystals I. Elechrolytic conduction in solid salts", *Transactions of the Faraday Society*, vol. 34, pp. 485–499, 1938.

[PAC 63] PACAULT A., *Eléments de thermodynamique statistique*, Masson, Paris, 1963.

[PRI 46] PRIGOGINE I., DEFAY R., *Thermodynamique chimique*, Dunod, Paris, 1946.

[PRI 66] PRINCE A., *Alloy Phase Equilibria*, Elsevier Publishing Company, Amsterdam, 1966.

[REI 87] REID R.C., PRAUSNITZ J.M., POLING B.E., *Properties of Gas and Liquids*, McGraw-Hill, New York, 1987.

[REN 71] RENON H., *Calcul sur ordinateur des équilibres liquide, vapeur et liquide-liquide*, Editions Technip, Paris, 1971.

[SOR 81] SORENSEN O.T., *Non-Stoichiometric Oxides*, Academic Press, New York, 1981.

[SOU 68] SOUCHAY P., *Thermodynamique chimique*, Masson, Paris, 1968.

[SOU 73] SOUSTELLE M., GARDET J.J., GUJLHOT B., "Diagrammes de phases et stœchiométrie", in SUCHET J.P. (ed.), *Diagrammes de phases et stœchiométrie*, Masson, Paris, 1973.

[SOU 90] SOUSTELLE M., *Modélisation macroscopique des transformations physico-chimiques*, Masson, Paris, 1990.

[SOU 93] SOUSTELLE M., "Défauts ponctuels dans les solides", in BERNACHE-ASSOLLANT D. (ed.), *Chimie, physique du frittage*, Hermes Science, Paris, 1993.

[SOU 10] SOUSTELLE M., *Handbook of Heterogeneous Kinetics*, ISTE Ltd, London and John Wiley & Sons, New York, 2010.

[SOU 11] SOUSTELLE M., *Cinétique chimique: éléments fondamentaux*, Hermes Science-Lavoisier, Paris, 2011.

[SOU 15a] SOUSTELLE M., *Phase Modeling Tools*, ISTE Ltd, London and John Wiley & Sons, New York, 2015.

[SOU 15b] SOUSTELLE M., *Modeling of Liquid Phases*, ISTE Ltd, London and John Wiley & Sons, New York, 2015.

[SYK 37] SYKES C., WILKINSON F., "The transformation in the beta brasses" *J. Inst. Metals*, vol. 61, p. 223, 1937.

[WHI 65] WHITE G.K., "The thermal expansion of alkali halides at low temperatures", *Proceedings of the Royal Society of London*, vol. 286, p. 204, 1965.

Index

A, B

C, D

E, F, G

H, I, J, L

K, L, M

N, O, P, Q

R, S, T

V, W, Y, Z

Other titles from

ISTE

in

Chemical Engineering

2015

SOUSTELLE Michel
Modeling of Liquid Phases

SOUSTELLE Michel
Phase Modeling Tools

2014

DAL PONT Jean-Pierre, AZZARO-PANTEL Catherine
New Approaches to the Process Industries: The Manufacturing Plant of the Future

2012

DAL PONT Jean-Pierre
Process Engineering and Industrial Management

2011

SOUSTELLE Michel
An Introduction to Chemical Kinetics

2010

SOUSTELLE Michel
Handbook of Heterogenous Kinetics

Lightning Source UK Ltd.
Milton Keynes UK
UKOW06n0115110915

258419UK00001B/2/P

9 781848 218666